CHEMISTRY
IN THE LAB

WILLIAM L. MASTERTON
University of Connecticut, Storrs, Connecticut

EMIL J. SLOWINSKI
Macalester College, St. Paul, Minnesota

EDWARD T. WALFORD
Cheyenne Mountain High School, Colorado Springs, Colorado

HOLT, RINEHART AND WINSTON, PUBLISHERS
New York • Toronto • London • Sydney

PREFACE

The experiments in this laboratory manual have been designed to go with the text "Chemistry" by Masterton, Slowinski, and Walford. There are one or two experiments for each chapter in the text, more than enough for a typical high school laboratory program.

In each experiment the authors have attempted to give the student sufficient information to allow him or her to understand how to perform and to interpret the experiment. Most of the experiments illustrate a theoretical or practical principle of chemistry. Many of them have unknowns with a property the student is asked to determine. The overall level of the experiments is compatible, we believe, with the text. An average student who applies reasonable effort to the task should be able to understand and profit from performing the experiments.

The experiments were selected both for interest and for their relevance to the concept being investigated. Some experiments are directly related to practical matters. Early in the course students prepare a flavoring, a paint, and plaster. In organic chemistry we have experiments in which the students prepare a soap and examine its properties. They also synthesize mauve dye, as Perkin did accidentally as a youth of eighteen in 1856. Late in the course some common household chemicals, like vinegar and antacids, are analyzed for their active ingredients.

Many of the experiments require calculations, which are for the most part very simple and directly related to the principles being demonstrated. We have attempted to take into account the feelings of uncertainty many students have toward mathematics, and usually include the equations the student must use. We feel that too many students have been turned off by what they felt was the inherent difficulty and mystery of chemistry and have tried to convey the impression that chemistry is both interesting and understandable.

Although some of the experiments are classic ones, a substantial number appear for the first time in this manual, at least as far as we know. Among these are an experiment designed to help students understand the mole, one on the heat of fusion of a solid, one on understanding electron charge clouds, and one on trace analysis. For the most part the equipment that is required is minimal and readily available. In two experiments absorption spectrophotometers are optional; in another, vacuum tube voltmeters would be useful though not essential.

The authors would like to acknowledge the assistance of Mr. Jon Thompson of Mitchell High School and of the students at Cheyenne Mountain High School whose classes tested the experiments over a period of two years.

We are also grateful for the help furnished by Katherine Arneson and Jeffrey Benson, students at Macalester College, who performed and criticized the experiments and did the typing of much of the manuscript. We hope the users of this manual, teachers and students alike, will feel free to send us their comments, suggestions, and criticisms regarding the manual. We assure you that we will acknowledge and carefully consider any suggestions that our readers would like to make.

INTRODUCTION TO THE LABORATORY

One of the first things a student beginning a study of chemistry must do is become familiar with the pieces of equipment and apparatus with which he or she will be working. Some items of chemical equipment will have obvious names and purposes, others are not so obvious. There are many pieces of equipment used in the laboratory, and in Figure 1 we have shown some of the more common ones. Learn the names of the items that are new to you, so that when they are used in an experiment, you will know which item to select.

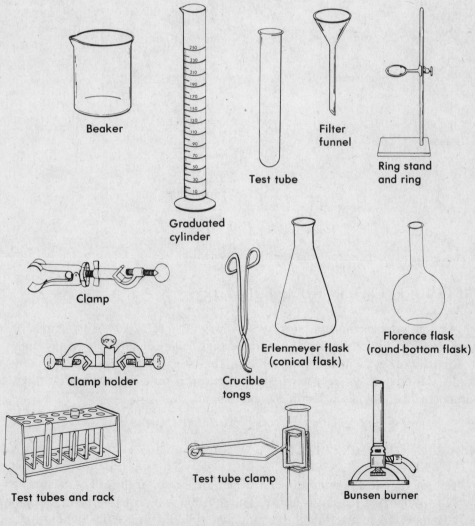

FIGURE 1 Chemistry laboratory equipment.

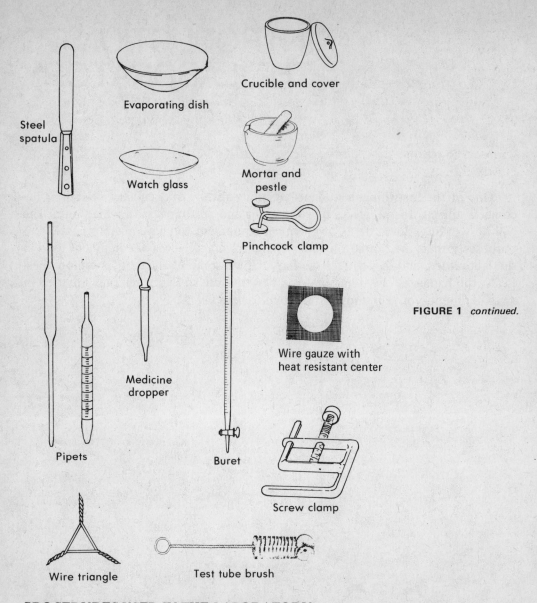

FIGURE 1 *continued.*

PROCEDURES USED IN THE LABORATORY

As you work in the laboratory you will learn to carry out many simple operations designed to accomplish the changes that chemical substances may undergo. In this section we will describe some of the basic procedures you will be using during this course. Where a particular operation is used in only one or a few experiments, it will be discussed in the experiment.

MEASURING OUT CHEMICALS

One of the most common operations in the laboratory involves measuring out a given volume of a reagent solution. The solution will probably be in a stock bottle and must be transferred to a graduated cylinder to get the right volume. In

Figure 2(A) we show the proper procedure. Note that you hold the stopper between your fingers while pouring, and you maintain contact between the bottle and the graduate. Pour a bit more liquid than you need into the graduate. Read the level in the graduate as shown in Figure 2(B), keeping a horizontal line of sight, and reading the bottom of the meniscus. Knowing that reading, pour out a little less than the amount of liquid you need into the beaker or test tube, and check the level in the cylinder. Then pour the rest of the required amount, slowly, so you don't get too much. When the level reaches the proper value, pour any excess reagent liquid into the sink or waste container, not back into the reagent bottle. (Why not? Because we don't want to contaminate reagents.)

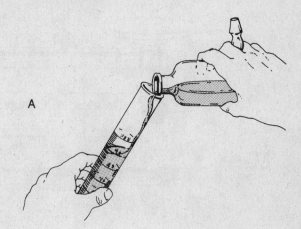

A

FIGURE 2 Measuring out a given volume of liquid.

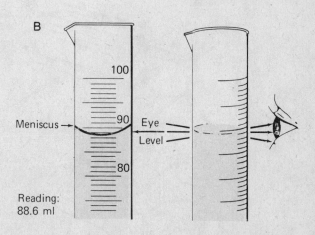

B

Meniscus →

Eye Level

100

90

80

Reading: 88.6 ml

If you need to weigh out a given mass of sample, you will need to use a balance. There are several kinds of balances in use in chemical laboratories; two of the more common ones are shown in Figure 3. The double pan balance at the top is used for rough weighings, and is good to about 0.1 gram. The triple beam balance shown below weighs to 0.01 gram. To operate either balance, you first check to find the balance point with zero load. Set all weights to the zero settings, release the beam, and gently touch the pan to set the beam in motion. Observe the swings of the beam, and note the scale readings at the extremes of the swings (Figure

$3(c)$). Take the average of the swing positions (a reading to the left of 0 is negative); that will be the rest point of the balance. Put the container for the solid on
the balance pan (never put the solid directly on the pan), and weigh the container
by moving the sliding masses until the rest point of the balance with load is the
same as it was without load. The sum of the masses you added is equal to the
mass of the container. The container to be used depends on the experiment. A
beaker, a watch glass, or a piece of filter paper make suitable containers.

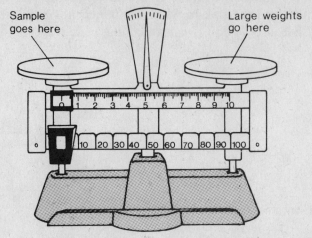

Sample goes here

Large weights go here

(a) Platform balance – weighs to 0.1 g

FIGURE 3 Using the balance.

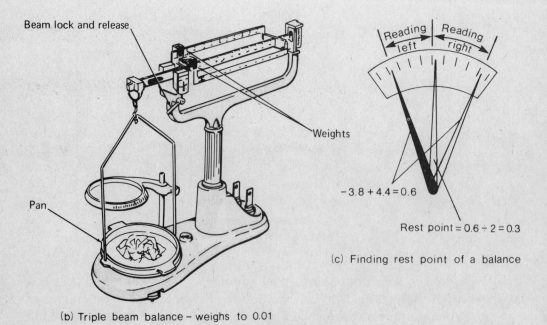

Beam lock and release

Weights

Pan

(b) Triple beam balance – weighs to 0.01

Reading left　Reading right

$-3.8 + 4.4 = 0.6$

Rest point $= 0.6 \div 2 = 0.3$

(c) Finding rest point of a balance

　　　To measure out the solid sample, bring the stock bottle to the balance. Set the
sliding masses on the balance to the proper value, taking due account of the mass
of the container. With a clean spatula, scoop out a small sample of solid. Put the
spatula over the container on the balance pan and gently tap the spatula until the

proper amount of solid falls into the container (Figure 4). If you need more solid, scoop out more sample. Put any excess into a waste container, not in the sink or in the stock bottle. If you tap off too much, pick up some solid from the container with the spatula and tap again.

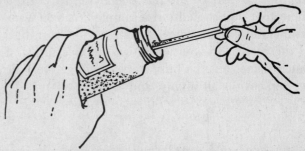

Scoop out a little of the sample with the spatula

FIGURE 4 Measuring out a solid sample.

Tap the spatula until the desired amount falls off

In working with small amounts of samples, where the actual amount is not important, it is often unnecessary to measure the actual amount. Rather, one makes an educated guess, and uses that amount. To establish an educated guess as to what constitutes one milliliter (1 ml, 1 cm^3), take a medicine dropper and add 1 ml of water to a 10 ml graduated cylinder. Pour the sample into a small test tube, and note the level. Measure out one more milliliter into the graduate, and pour that into the test tube. Remember where the level is for 1 ml and 2 ml, and from then on fill to that level to obtain those volumes. With small masses, the same kind of approach is useful. Weigh out 1 gram of a typical solid, say some common table salt, and pour that sample into a small test tube. Note the level in the tube, and when you need a gram of solid, fill the tube to that level.

HEATING SAMPLES

In almost every experiment, it is necessary to heat a liquid or a solid sample. This is done in different ways, depending on the sample. In nearly all cases, however, you will use a Bunsen burner to furnish the heat. In Figure 5, we show a typical Bunsen burner. To light the burner, open the gas jet and ignite the gas with a spark igniter or a match. The burner flame should be blue or violet, not yellow. Using the air adjustment, increase or decrease the amount of air until the flame is nonluminous and nearly silent. You can control the amount of gas with the gas jet, or, on some burners, with a control on the bottom of the burner. The hottest part of the flame is at the top of the inner light blue cone, and it is there that you should put a sample that you want to bring to the highest possible temperature. Sometimes, a burner will "burn back", in which case the gas ignites at the base of the burner. It will make a rather strange noise, and there won't be much flame, but the burner itself will get very hot. If that happens, turn off the burner, cut the air supply, and re-ignite.

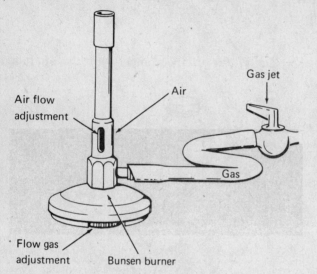

FIGURE 5   The Bunsen burner.

If you wish to heat a liquid sample, but do not want to boil it, you should use a hot water bath like that shown in Figure 6. To make a water bath you need a beaker (250 ml is a good size to use); fill the beaker about ¾ full of water and set it on a piece of wire gauze supported by an iron ring. Heat the water with the burner to the desired temperature; keep the water at that temperature by judicious adjustment of the burner. A test tube containing the liquid sample will quickly come to temperature when the tube is put in the hot water bath.

To boil a liquid sample, we use a beaker of the proper size, arranged as in Figure 6. Never try to boil a liquid in a test tube. What will happen if you do that is that the liquid will bump out, on you or your neighbor. So don't try it. When boiling a small volume of liquid you need to be careful not to heat too strongly. It is very easy to boil off all the liquid, at which point the remaining solid will get very hot and this may ruin the experiment or produce a toxic product, neither of which is at all desirable. An evaporating dish is used when you intend to evaporate a solution to dryness.

To heat a solid we use a crucible, which is made of porcelain and able to tolerate high temperatures better than glass. The crucible is supported on a clay triangle, which in turn is supported by an iron ring. The crucible can stand being heated to red heat, which is necessary in some experiments. To cool a crucible, either let it cool on the clay triangle, or lift it with tongs and put it on an asbestos cement pad. Don't put a hot crucible directly on the lab bench or on a balance pan.

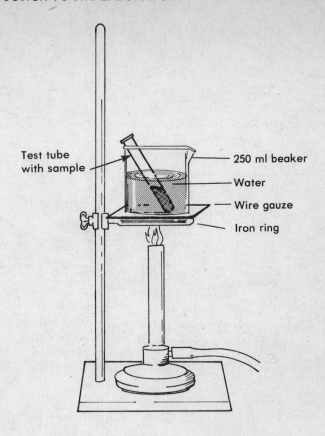

Test tube
with sample

250 ml beaker

Water

Wire gauze

Iron ring

FIGURE 6 **Heating a liquid sample
in a water bath.**

SEPARATIONS

Frequently it is necessary to separate a solid from a liquid with which it is mixed. Depending on the circumstances, this is done in various ways. The simplest is to let the solid settle out, and then pour off, or decant, the liquid. To obtain a somewhat better separation when small amounts of material are involved, one often uses a centrifuge; the centrifuge has holders that will take small test tubes. In operation the test tubes are spun rapidly, and any solids present go to the bottom of the tubes. The liquid can then be decanted and so separated from the solid.

Perhaps the most common method for separating a solid from a liquid is filtration. In Figure 7 we have shown a filtration setup. A piece of filter paper is folded in half, and then in half again. A small tear is made on one corner of the paper, and then the filter paper is put into the funnel with the tear on the outside. The paper is moistened with distilled water and pressed against the funnel wall. The mixture to be filtered is added to the funnel, decanting as much liquid as possible. The solid is poured in with the last of the liquid. A stirring rod is used to move as much of the remaining solid as possible into the funnel. The last of the solid is washed out of the beaker into the funnel with a stream of distilled water from a wash bottle.

LABORATORY REPORTS

In each experiment you perform you will make observations and obtain data. This information, along with calculations and other conclusions to be drawn from

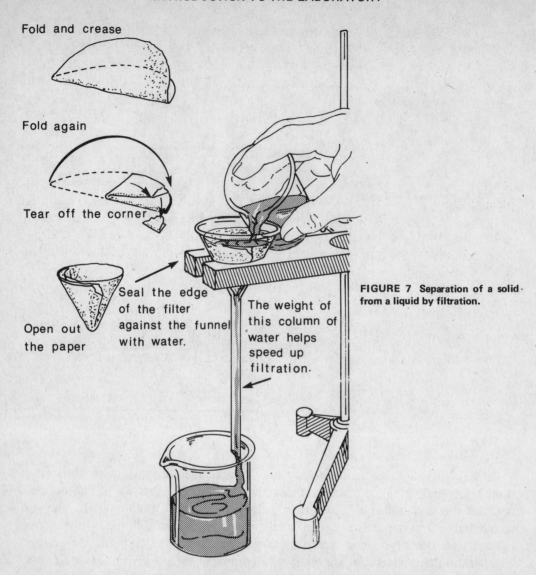

Fold and crease

Fold again

Tear off the corner

Open out the paper

Seal the edge of the filter against the funnel with water.

The weight of this column of water helps speed up filtration.

FIGURE 7 Separation of a solid from a liquid by filtration.

the experiment, should be entered in your laboratory notebook. Your teacher will tell you what sort of notebook to use and how to prepare your reports. You should bring your notebook to every laboratory session, along with your laboratory manual and calculator.

Before coming to laboratory, review the chapter in the text that is mentioned in the Preliminary Study section. Answer as best you can any questions that are asked in that section. Read over the experiment so that you have some idea what you will be doing.

During the laboratory period, or, preferably, in advance, you should prepare a Data Table for the experiment being performed. With most experiments, a Sample Data Table has been included, to give you an idea about how to arrange the data you obtain. The Sample Data Table is usually incomplete, so you should add to it as necessary to allow room for all the data you collect during the experiment. Enter data and observations in the Table at the time you obtain such information.

Following the Data Table, carry out the calculations and answer the questions that appear in the Calculations and Questions section. It is best if you can complete this part of the assignment while you are still in the laboratory, but if you have to do it outside of class, make sure that your Data Table is complete before leaving the laboratory.

GENERAL

The experiments in this laboratory manual have been designed to illustrate the ideas that are presented in the text. Performing experiments and figuring out what they mean can be very helpful to you in understanding chemistry. Chemistry may not be easy for you, but if you give it your best effort you will find that you will do better than you think you might. Many of you will enjoy it and, who knows, some of you might even find out that you would like to be a chemist. We authors will be glad to hear from you if that turns out to be the case. In fact, we'd be glad to hear from you even if it doesn't.

SAFETY IN THE LABORATORY

A chemistry laboratory can be, and should be, a safe place in which to work. However, as in the kitchen or workshop in your home, there are hazards to be recognized and avoided. Chemicals, if dealt with casually or without knowledge of their properties, can behave in unexpected ways. Laboratory burners can cause dangerous fires if used improperly. Glassware can burn you if it is hot, and cut you if it is broken. Accidents in the laboratory are not necessary, and can almost always be prevented if those in the laboratory are aware of the possible dangers and behave intelligently.

When you come to the laboratory, dress sensibly. If you have long hair, confine it to minimize the chance that it will get near a flame or get caught in a piece of moving apparatus. Wear shoes that protect your feet adequately. If you have a lab apron or coat available, it makes sense to wear it to protect your clothing.

In the laboratory the single most important thing you can do to protect yourself is to wear safety goggles. Your eyes are vulnerable to chemical splashes and hot sparks, so you must keep your eyes covered. Do not wear contact lenses in the laboratory. Get safety goggles from the stockroom. Wear them whenever you are in the laboratory.

The most common accident in the beginning chemistry laboratory occurs when a student tries to insert glass tubing, a thermometer, or a glass rod into a hole in a rubber stopper. The glass breaks because the student subjects it to excess force in the wrong direction, and sharp glass cuts the student's finger or hand, sometimes severely. Such accidents are completely unnecessary. If you need to work with glass in this way, ask your teacher how to do it properly before you try it. If you do cut yourself, go to your teacher and he or she will decide on the proper treatment and report to be filed.

Although other kinds of accidents are less common, they do occasionally occur, and should be considered as possibilities. Burns may be caused by touching hot glassware or iron rings. A volatile organic liquid may ignite if it gets too close to a Bunsen burner flame. You may spill a caustic reagent on yourself or your neighbor, or somehow get some chemical into your eye or mouth. An all too common response in such a situation is panic. A student may do something irrational, like run from the laboratory when the remedy for the accident is close at hand. If an accident happens to another student, watch for signs of panic and tell the student what to do, and if necessary help him or her to do it. Call your teacher for assistance. Chemical spills are best handled by quickly washing the area with water from the nearest sink. Severe spills may require using an emergency shower or eyewash and removing affected clothing. If a fire occurs, it can be put out with a fire extinguisher, a blanket, or water as seems most appropriate. If the fire is on a lab bench or in a beaker and does not appear to require immediate action, have your teacher put the fire out. In any event, know where the safety equipment is located in the laboratory, so you can find it in case you need it.

In the laboratory do not touch or taste any chemicals unless specifically directed to do so. Avoid exposure to vapors given off by volatile chemicals or by chemical reactions; if necessary, work in the hood or open the windows in the lab. If you are directed to smell a vapor, sniff cautiously, at least until you know what the odor is. When you are finished working in the laboratory, it is a good idea to wash your face and hands.

In this laboratory manual we have attempted to describe experiments which are safe when performed according to the procedures we have given. Follow directions when carrying out any experiments and observe any indicated precautions. Do not underestimate the power of any chemicals, and above all do not perform any unauthorized experiments.

Every chemist who has worked in the field for a few years has seen accidents or near accidents happen, to himself or to others. Chemical accidents are never enjoyable. Do what you can to prevent them, in this course and in any future work you may do.

CONTENTS

LABORATORY TECHNIQUES: MAKING MEASUREMENTS

OBJECTIVES

1. To practice the techniques of making mass, length, and volume measurements.
2. To become familiar with the operation of laboratory balances.
3. To indirectly determine the thickness of a metal foil.
4. To identify an unknown metal by determining its density.
5. To determine the precision error and accuracy error of a measurement.
6. To become familiar with the metric system and metric-English conversions.
7. To practice the correct use of significant figures.

DISCUSSION

Experiments in chemistry require a variety of instruments and techniques for making measurements. Some of the more basic procedures are described in the Introduction (p. vi). Those which are used less frequently will be described within an experiment as they are needed. In particular, the techniques of measuring mass, volume, and temperature will be used throughout this course. This experiment provides an opportunity to practice some of these techniques while, at the same time, gathering useful information.

Measurement Errors: All measurements involve two types of error, precision and accuracy. The precision error is related to the reproducibility of the measurement. Consider an object weighed by four students on a platform (double pan) balance. The results obtained are: 9.8 g, 10.2 g, 10.0 g, 10.4 g. The average mass is found by adding the measurements and dividing by their number.

$$\overline{m} = \frac{9.8 + 10.2 + 10.0 + 10.4}{4} = 10.1 \text{ g}$$

The uncertainty (or precision error) in any mass measurement using this balance is equal to the average error of the measurements. An error is calculated as the absolute difference between the measurement and the average of many measurements. In this case, the individual errors and the average error are:

Reading	Error
9.8	0.3
10.2	0.1
10.0	0.1
10.4	0.3
	0.8 Total errors

$$\text{Average Error} = \frac{0.8}{4} = \pm 0.2$$
(Precision Error)

The uncertainty is a guide to the proper use of an instrument. If the error is large compared to the size of the measurement a more sensitive instrument should be used. For example, it would be meaningless to weigh objects of 0.1 g or less on a balance whose average error is 0.2 g. The centigram (triple beam) balance whose average error is in the neighborhood of ± 0.01 g would be a better choice. In most laboratory work, precision errors are taken into account by recording all data to the proper number of significant figures. For example, in using the platform balance all measurements should be recorded to the nearest 0.1 g while those made with a centigram balance should be recorded to the nearest 0.01 g.

The accuracy error is related to the deviation of the measured value from the true value. Measurements will differ from the true value, owing to consistent errors in the instrument, the particular technique being used, or to some quirk of the experimenter. In many experiments you will be asked to calculate an accuracy error based on percentage deviation from a known value. This is accomplished by using the relationship:

$$\% \text{ Accuracy Error} = \frac{\text{True Value} - \text{Measured Value}}{\text{True Value}} \times 100$$

Assuming that the mass of the object above was measured by the National Bureau of Standards and was found to be 10.4 g, the percent accuracy error is:

$$\% \text{ Accuracy Error} = \frac{(10.4 - 10.1)}{10.4} \times 100 = 3\%$$

Extensive and Intensive Properties: Some properties of a sample vary with the amount of sample. The mass and volume, for example, both increase with the amount. They both double if the amount doubles; a fractional increase in amount will result in the same fractional increase in the mass and volume. Properties like mass and volume are said to be extensive properties. Extensive properties are proportional to the amount of sample. Cost is another example of an extensive property.

There are other properties of substances which do not vary with amount. Many of these properties are based on a certain amount of material, say 1 gram or 1 cm^3, so it doesn't make any difference how much we have; we only consider a fixed amount when we establish the property. For example, if we measure the mass and the volume of a sample, and then divide the mass by the volume, we get a property of the sample, which we call density, that does not depend on how big the sample is:

$$\text{density} = \frac{\text{mass of sample}}{\text{volume of sample}} \tag{1.1}$$

The density of a substance, unlike its mass or volume, is characteristic of the nature of the substance. A substance like gold will have a particular density, different from that of iron, or water, or lead. We can, and do, use density measurements to help us identify substances. Properties like density, which do not depend on amount of substance, are called intensive properties. There are many intensive properties, of which color, hardness, and melting point are some examples. Sometimes, if two substances have similar densities, it is necessary to measure one or more other intensive properties in order to establish which substance is actually in an unknown sample.

Density Measurements: To measure the density of a substance, we need to measure the mass and volume of a sample of that substance. The mass of a substance is measured on a balance. Its volume can be measured in various ways. If the substance is a liquid we can determine its volume with a graduated cylinder or other container which is calibrated for volume. We will find out more about this method in Experiment 3. If the substance is a solid with a regular shape, such as a cube or a cylindrical rod, we can obtain its volume by direct measurement of its dimensions. If the solid consists of a powder, or crystals, or irregular chunks, the volume can often be measured by the liquid displacement method. To a known volume of water in a calibrated container such as a graduated cylinder, one adds a known mass of solid sample. The water level rises, by an amount equal to the volume of the solid that was added. The volume of the solid is the difference in the volumes indicated by the final and initial water levels.

$$\text{Volume of solid} = \text{Final volume} - \text{Initial volume} \tag{1.2}$$

In this experiment, we will carry out several measurements that involve density. After you have learned to work with the balance, you will be given a rectangular piece of metal foil of known density. By measuring the mass of the foil, you can calculate its volume by using its density:

$$\text{Since} \quad \text{density} = \frac{\text{mass}}{\text{volume}} \qquad\qquad \text{volume} = \frac{\text{mass}}{\text{density}} \tag{1.3}$$

From the dimensions of the foil you can find its area:

$$\text{area} = \text{length} \times \text{width} \tag{1.4}$$

The volume of the foil is equal to its area times its thickness:

$$\text{volume} = \text{area} \times \text{thickness} \tag{1.5}$$

Since you found the volume from the mass and density, and also know the area, it is an easy calculation to establish the thickness of the foil. The experiment shows how, given a property of a sample, like density, one can calculate another property, like thickness, that might otherwise be difficult to measure.

In another part of the experiment you will determine the density of a sample of unknown metal. Here we will measure the mass, as always, on a balance, and will find the volume by the liquid displacement method.

Finally, you and other students will measure the mass of an assigned object. From the different masses obtained, we will find the average mass of the object and the precision error in the experiment.

TABLE 1–1   The Densities of Some Common Metals

METAL	DENSITY (g/cm^3)	METAL	DENSITY (g/cm^3)
Aluminum	2.70	Lead	11.35
Cadmium	8.65	Magnesium	1.74
Chromium	7.20	Silver	10.5
Cobalt	8.9	Tin	7.28
Copper	8.92	Zinc	7.13
Iron	7.86		

PRELIMINARY STUDY

1. Read "Introduction to the Laboratory" in the Introduction of this manual.
2. Review Sections 1.3 and 1.4 in your text.
3. Practice Problem: Using an automobile odometer, a student measured the distance between two crossroads as 13.6 km. A survey map showed the actual distance to be 12.5 km. Calculate the percent error in the odometer measurement. (Ans.: 8.8%)

PROCEDURE

1. Operation of the Balance

Take a few minutes to become familiar with the operation of your laboratory balance. Make sure that the balance is level and adjust its rest point to zero. Then make a few practice weighings and have someone check your readings.

2. Measurement of the Thickness of a Metal Foil

Obtain a rectangular piece of metal foil from your instructor. You will be told the identity of the foil and its density. Measure the mass, length, and width of the foil as precisely as your equipment will allow. Your instructor will tell you the precision you should try to obtain.

3. Measurement of the Density of a Metal

Obtain a sample of an unknown metal from your instructor. Record the unknown number. Weigh the sample on the balance. If the sample is in a container, weigh the container and its contents. Then pour the sample into a dry beaker and weigh the container. Partially fill (over the 15 cm³ mark) a 50 cm³ graduated cylinder with water. Measure and record the water level. Carefully, without splashing out any water, slide the metal into the water in the graduate and record the new volume. The water must completely cover the metal (why?).

4. Measurement of Precision Error

Weigh an object assigned to you and record its mass on the blackboard. The same object should be weighed by other students and their results similarly recorded.

SAMPLE DATA TABLE   **EXPERIMENT 1**

Thickness of a Metal Foil

metal _____   density _____ g/cm³   mass _____ g

length _____ cm   width _____ cm

Density of a Metal

unknown number _____   vol. of water (initial) _____ cm³

mass of container plus metal _____ g   vol. of water (final) _____ cm³

mass of container _____ g   *Precision of Balance*

mass of metal _____ g   mass of class object _____ g

CALCULATIONS AND QUESTIONS

In all calculations, observe the rules on significant figures given in Section 1.4 of the text.

1. (a) Calculate the area of the foil used in Procedure 2.
   (b) Using the density and mass of the foil, determine its volume.
   (c) What is the thickness of the foil?
   (d) Convert the thickness of the foil from centimeters to inches. Use the conversion factor method described in Section 1.3 of the text. Note that 1 inch = 2.54 cm, so $\dfrac{1 \text{ inch}}{2.54 \text{ cm}} = 1$. If it is not clear how to proceed from here, see Example 1.3 in the text.
2. (a) Calculate the volume of the metal used in Procedure 3.
   (b) Determine the density of the metal from its mass and volume.
   (c) Attempt to identify your metal by comparing its density to those listed in Table 1–1. Based upon your identification, calculate the percent accuracy error in the density measurement.
   (d) Convert the density of the metal from g/cm^3 to $pounds/ft^3$. The following equations can be used to obtain the needed conversion factors:

$$1 \text{ pound} = 453.6 \text{ g} \qquad 1 \text{ foot} = 12 \text{ inches} \qquad 1 \text{ inch} = 2.54 \text{ cm}$$

$$(1 \text{ foot})^3 = (12)^3 \text{ inches}^3 \qquad\qquad (1 \text{ inch})^3 = (2.54)^3 \text{ cm}^3$$

   If it is not clear how to complete the conversion, see Example 1.4 in the text.
3. (a) Use the class data to determine the average mass of the object weighed in Procedure 4.
   (b) Find the precision error in the measurement of the mass. To how many significant figures can the mass be properly reported?

EXTENSIONS

1. Obtain a micrometer and measure the thickness of the metal foil in Procedure 2. Calculate the percent accuracy error of your thickness determination based on the micrometer measurement.
2. If the metal piece used in Procedure 3 has a regular shape, measure its dimensions and determine its volume mathematically. Calculate the density using this volume and compare your answer to that obtained by water displacement. Which method was more accurate?
3. Partially fill a 50 cm^3 graduated cylinder with water so that the meniscus (water level) is not even with a mark. Have at least ten students make independent readings of the volume. The volume should be estimated to the nearest 0.1 cm^3. Calculate the average volume and the precision error of the measurement.

LABORATORY TECHNIQUES: MAKING CHEMICAL PRODUCTS

OBJECTIVES

1. To gain experience in using basic laboratory techniques.
2. To develop your laboratory skills by preparing some products of chemistry.
3. To observe the properties of several chemical substances.

DISCUSSION

Work in the chemistry laboratory involves many different procedures and techniques. In the previous experiment you learned how we measure mass and one of the ways we measure volume. In this experiment you will be introduced to several techniques for carrying out chemical reactions and separations. You should learn how to operate a Bunsen burner, how to heat a liquid in a test tube, how to filter and wash a precipitate, and how to transfer chemical substances from one container to another. We will use these operations to make some chemical substances with which you are familiar.

In the Introduction to this manual we have discussed the procedures we have mentioned, along with several others, showing the apparatus that is involved and describing the proper way to carry out the procedure. When, in this experiment and others that follow, you encounter a procedure with which you are not familiar, refer to the Introduction before you attempt to do that part of the experiment. You will feel more sure of what to do and will probably do a better job. In some experiments where a somewhat specialized procedure is used we will include a description of the procedure in the experiment, but for the most part they are discussed in the Introduction.

In the Introduction we also have diagrams of the pieces of laboratory equipment you will be using. Consult that part of the Introduction whenever you are in doubt as to what a particular item looks like or how it is used.

A chemistry laboratory, above all, should be a safe place in which to work.

Chemicals, however, can behave in unexpected ways, so it is important that you always follow instructions. Above all, do not perform any unauthorized experiments. In experiments involving chemical reactions, you should wear your safety glasses at all times. If particular chemicals or procedures are potentially hazardous, we will include a special caution statement where they are used. In the Safety in the Laboratory section at the beginning of this manual we discuss how to safely carry out such procedures as smelling a vapor and handling hot equipment, as well as how to deal with accidents that may occur. Please read that section and become familiar with it. Most accidents in the chemistry laboratory can be avoided. Let us avoid them, by learning procedures that are safe and using them.

PRELIMINARY STUDY

1. Review the procedures used in the laboratory in the Introduction for:
   (a) transferring chemicals.
   (b) the operation and adjustment of the Bunsen burner.
   (c) the separation of mixtures by filtration.
2. Refer to Figure 1 in the Introduction and identify the following pieces of apparatus:

beaker	mortar and pestle
iron ring	spatula
ring stand	graduated cylinder
wire gauze	filter funnel
crucible tongs	test tube

PROCEDURE

Wear your safety goggles, apron, and gloves during this experiment.

1. **Flavors and Perfumes.** Flavors are detected by our sense of taste and perfumes by our sense of smell. Originally most of these products were extracted from natural sources. Today the active ingredients in flavors and perfumes have been identified and many of these can be produced in the laboratory. One such substance, which you may be able to identify, can be made by the following procedure.
   (a) Place 0.5 g of salicylic acid in a clean dry test tube and dissolve the solid in 3 ml of methyl alcohol.
   (b) After all of the solid has dissolved, slowly add 10 drops of concentrated sulfuric acid. Gently swirl the contents of the tube after the addition of each drop.

   CAUTION: Concentrated sulfuric acid is very harmful to the skin. Be very careful when using this acid. If you should get any on your skin or clothing, wash it off immediately with plenty of water. Any acid spills on the desk top should be removed with a damp sponge.

   (c) Heat a 250 ml beaker two-thirds full of water to 60–70°C. Place the tube and reactants in the water and keep at that temperature for 5 minutes. Remove the tube from the water bath, and pour the contents into the water in the beaker. Note the odor. The pleasant smell is caused by the presence of methyl salicylate produced in the reaction. (After the solution cools, unreacted salicylic acid may precipitate out.)

2. **Moisturizing Cream.** Early cosmetics were made from commonly available materials. The colonists of America used mutton tallow to deal with the problem of rough hands. Today we rarely use mutton tallow for chapped hands, since chemists have found that other blends of substances can produce better results. The recipes of an earlier time have been replaced by more uniform procedures that consistently produce a desired product.

   (a) Mix 0.1 g of potassium carbonate with 1 ml of glycerol and 10 ml of distilled water in a 100 ml beaker. Heat this mixture to about 80°C.

   (b) Heat about 3 g of stearic acid in a test tube until it just melts.

   (c) Pour the molten stearic acid into the hot solution in the beaker and stir the creamy mixture rapidly until it thickens. If the moisturizing cream does not reach a solid paste-like consistency, heat it slightly to just boiling and then cool it in an ice bath.

   (d) When the cream has cooled, feel the texture of the cream with your fingers. Rub a small portion on your hand. How does it compare to commercial preparations?

3. **Paint.** Paints have three components; a binder, a solvent, and a pigment. The binder is a film-forming ingredient, usually a drying oil or some plastic-like substance. Linseed oil acted as the binder in early paints. The paint was difficult to work with and required several days to dry.

   (a) Weigh 0.3 g of potassium chromate and 0.5 g of lead nitrate on separate pre-weighted pieces of filter paper. Add 3 ml distilled water to each of two test tubes. Dissolve the potassium chromate in the water in one of the tubes and the lead nitrate in the other. Pour one solution into the other and note the precipitate of lead chromate which is formed. The lead chromate is the pigment we will use in our paint.

   (b) Set up a funnel for filtration. Filter and wash the precipitate on the filter paper several times with distilled water from a wash bottle.

   (c) When the filtration is complete, open up the filter paper and place it in a drying oven to dry for a few minutes. If a drying oven is not available, transfer the precipitate with a spatula to a piece of dry filter paper. Spread the precipitate evenly over the center of the paper to allow the paper to soak up as much water as possible. Then wash the precipitate several times with a fine stream of acetone from a wash bottle. The acetone evaporates rapidly and will speed up the drying process. Let the precipitate dry until it is easily crumbled to a fine powder.

   CAUTION: Acetone is very flammable. Never use it near an open flame.

   (d) Mix a small amount of the powdered lead chromate with a few drops of linseed oil and turpentine until a paint-like consistency is achieved. Use some of the paint on a wood splint. Examine the splint when the paint is dry.

CAUTION: Lead compounds are poisonous and are no longer used in commercial paints. Wash your hands after completion of this part of the experiment.

4. **Plaster.** Gypsum is a mineral that occurs in large deposits throughout the world. It is a hydrated form of calcium sulfate. Hydrates contain water which can be removed by heating. When gypsum is partially dehydrated, a product known as "plaster of Paris" is obtained. If water is added, the dehydration reaction is reversed. The product is another form of gypsum which is hard, reasonably strong, and used for making casts, plaster, and wallboard.

   (a) Weigh out a 10 g sample of gypsum and place in an evaporating dish. Put the dish on a wire gauze on a ring stand over a Bunsen burner. Heat the dish strongly for 10–15 minutes. Using crucible tongs, place the dish on a heat resistant pad to cool.

   (b) While the dish is cooling, prepare a coin or key by coating it with a thin film of oil.

   (c) When the evaporating dish has cooled, pour the product, which is plaster of Paris, onto a glass plate. Add just enough water, while stirring, to make a thick paste.

   (d) Quickly press the metal object into the surface of the paste and allow it to remain until the paste hardens. Remove the metal object and inspect the plaster mold.

QUESTIONS

1. Describe the properties of the chemical products which you made. Be sure to include the properties which give them commercial value.
2. Write an advertisement for any one of the products which you made.
3. Describe the proper technique for:
   (a) testing the odor of a substance.
   (b) heating a liquid in a test tube.
   (c) handling a hot beaker or evaporating dish.
4. What technique should be used to make the following separations:
   (a) mud from muddy water?
   (b) salt from ocean water?
5. Explain the purpose of the controls on a Bunsen burner. What condition causes the burner to produce a luminous flame?

EXTENSIONS

1. Explain the hazards involved in:
   (a) stirring a solution with a thermometer.
   (b) evaporating a salt solution to dryness.
   (c) leaving a burner unattended while heating a solution.
   (d) spilling a solution and not wiping it up.

(e) discarding waste solids in a sink.
(f) returning unused solutions to their stock bottles.
(g) drinking water from a laboratory beaker.
(h) failing to wear protective eyeglasses while in the lab.

RELATIONSHIPS BETWEEN VARIABLES: MASS AND VOLUME OF A LIQUID

OBJECTIVES

1. To determine the relationship which exists between the mass and volume of a liquid.
2. To find a mathematical expression for the relationship.
3. To use the relationship for prediction.
4. To practice making graphs.

DISCUSSION

The study of a particular phenomenon often suggests that two measured properties are related to each other. An experiment can then be designed to determine this relationship by measuring the effect that changing one property (the independent variable) has on the other (the dependent variable). In this experiment we will study the way the mass and the volume of a liquid are related.

In searching for relationships between two properties it is frequently useful to make a graph showing how one property varies with the other. Each of the coordinates on the graph represents one of the properties being studied. In an experiment involved with mass and volume we might have the volume values along the x-axis and the mass values along the y-axis. To make the graph we would put data points for each sample at the proper values of volume and mass. If, for example, we find that a sample containing 2.0 cm³ of liquid has a mass equal to 2.80 grams, we would put the data point for the sample at the point on the graph network where volume equals 2.0 cm³ and mass equals 2.80 grams. By using all the data points we get an idea of how the mass varies with volume. In order to be able to say what the mass of the sample would be for volumes other than those we measured, we draw a line through the data points which best fits the trend shown by those points. Having found the line showing how two properties like mass and volume are related, it is often possible to find a mathematical equation for that line. That equation can then be used to calculate one property when the other one is known. It can sometimes also be used to establish an intensive property of the substance being studied. In this experiment we will be doing all of these things, with the ultimate purpose of understanding that property of a liquid called its density.

Our actual experiment is quite simple. We will measure the volume and mass of several different liquid samples, working first with water and then with an unknown liquid. These measurements will furnish us with the data from which we can make and interpret graphs for the mass-volume relationships of water and the unknown.

PRELIMINARY STUDY

1. Read the section on making graphs in the Appendix of your text.

PROCEDURE

1. Weigh a clean, dry 10 cm^3 graduated cylinder to the nearest 0.01 g.
2. Using a medicine dropper, add water until the level is as close as possible to the 2 cm^3 mark. Read the level and weigh again. Make the volume measurement at the bottom of the liquid meniscus.
3. Add another 2 cm^3 of water. Read the volume and reweigh the cylinder. Continue in this way to make volume and mass measurements for every two cubic centimeters up to the 10 cm^3 mark. Record all volumes to the nearest 0.1 cm^3 and all masses to the nearest 0.01 g.
4. Empty and dry the graduated cylinder. Repeat Procedures 2 and 3 with an unknown liquid. Record your values for the unknown liquid.

SAMPLE DATA TABLE   **EXPERIMENT 3**

Mass of graduated cylinder _____ g   Unknown liquid number _____

Total volume of water	Mass of grad. cylinder and water	Total mass of water
_____ cm³	_____ g	_____ g

Use a similar arrangement for data on the unknown liquid.

CALCULATIONS AND QUESTIONS

1. From the data obtained, calculate the total mass of the water in the cylinder for each of the volumes you measured. Carry out similar calculations for the total mass of the unknown liquid for each volume.

2. On graph paper plot the data you obtained for the water samples. Plot mass on the y-axis and volume on the x-axis. You should have a point on the graph paper showing the volume and mass for each sample you measured. When you have completed plotting the data, draw a straight line through the data points in such a way as to minimize the distances the points lie off the line. Your line should go through the origin. (Why?) Repeat the procedure, plotting the data you found for the unknown liquid, and drawing a line through those data points. Label the first line as WATER, and the other line as UNKNOWN LIQUID.

3. The lines on the graph describe how the volume and mass of each of the two liquids are related. For any given volume, we can find the mass that that volume of water, or unknown liquid, would have. Using the line for water on the graph, find the mass of water which would have the following volumes: 1.0 cm^3, 3.0 cm^3, 5.0 cm^3, 7.0 cm^3, 9.0 cm^3. Then, using the line for the unknown, find the mass of unknown liquid which would have each of these volumes.

4. By looking at the results just obtained from your graphs, it should be apparent how you could find the mass of *any* volume of water, or unknown liquid, that you might select. State in words how you would go about making the calculation.

5. It is possible to write the equivalent of your word statements in 4 in two mathematical equations, one for water and one for the unknown. See if you can write those equations. These equations relate mathematically the mass of the liquid to its volume, and so must contain both mass and volume as variables.

6. Each of the equations in 5 contains a constant term. What units must the constant term have if the units in the equation are to be consistent? Physically, what does each constant term tell us about the liquid to which it is related? What physical property of the liquid is equal to that constant term? Write a sentence in which you state the value of that property for water and for the unknown.

7. Using your equations in 5, or your word statements in 4, find the mass of 14 cm^3 of water. Find the mass of 24 cm^3 of the unknown.

8. Revising the equations you developed in 5, find the volume of water you would need to measure out to obtain 68 grams of water. What volume would be needed to furnish 29 grams of the unknown? Would you say there are advantages in measuring volume rather than mass to obtain a sample having a certain mass?

EXTENSIONS

1. In this experiment the only two variables we considered were mass and volume. Temperature might also be an important variable, and we ignored it. Suggest how you might carry out an experiment to determine whether temperature needs to be measured in experiments like the one you performed.

2. Write an equation expressing the relationship, if any, which exists between the listed variables. Sketch a graph consistent with the relationship.

(a) temperature in degrees Celsius and degrees Fahrenheit.

(b) a student's chemistry grade and that student's month of birth.

(c) the length and width of a rectangle of area equal to 100 cm^2.

3. Go to the library and obtain world population data covering an extended period. Plot population on the y-axis and the year on the x-axis. Describe the relationship between these variables. Predict the world's population in the year 2000.

COUNTING PARTICLES AND FINDING THEIR RELATIVE MASSES

OBJECTIVES

1. To develop and use a new system for counting small particles.
2. To determine the relative mass of different small particles by comparison with an arbitrarily chosen mass.
3. To better understand the mole system and the table of atomic masses by comparison with a model system.

DISCUSSION I: WAYS TO COUNT BEANS

Counting a large number of eggs, or sheets of paper, or beans, is a time-consuming, tedious task. One way to minimize the difficulty of such a job is to use units larger than "one" in the counting process. It's a lot easier to count 15 dozen eggs than it is to count 180 eggs. Two reams of paper mean the same to the stationery store manager as 1000 sheets, without counting. In chemistry, we have a system for counting atoms and molecules that is somewhat different from the dozen or the ream, but which has similar advantages. In order to help you become familiar with the chemical system we will do an experiment involving different kinds of beans. In the experiment, we will count beans in groups called "bunches." You will determine the number of beans in a bunch by relating it to an arbitrary standard of mass. As you will see, this procedure will allow you to avoid the burden of actually counting large numbers of beans. In the second part of the experiment, we will see how a bunch of beans is related to a mole of atoms or molecules.

PRELIMINARY STUDY

1. Review the concept of the mole and mole-gram conversions in Section 2.6 of your text.

17

PROCEDURE

1. (a) Weigh an empty paper cup (or any other suitable container). Count out exactly 100 beans of one type, discarding any beans which differ greatly from an average bean. Weigh the cup of beans.
    (b) Repeat this procedure for each type of bean provided.
2. (a) Calculate the average mass of one bean of each type and record it in the data table.
    (b) Determine the relative mass of each type of bean by comparison to the lightest type of bean. Record these values in the data table.

$$\text{Relative mass} = \frac{\text{Avg mass of bean}}{\text{Avg mass of lightest bean}}$$

3. Weigh out the relative mass (in grams) of each kind of bean and count the beans weighed.

SAMPLE DATA TABLE				EXPERIMENT 4
	Bean 1	Bean 2	Bean 3	Bean 4
Mass of beans and cup	_____ g			
Mass of cup	_____ g			
Mass of beans	_____ g			
Average mass	_____ g			
Relative mass	_____			
No. of beans in relative mass	_____			

CALCULATIONS AND QUESTIONS: I

1. (a) If a "bunch" is defined as the number of beans in 1.00 g of the lightest bean, how many beans are in a bunch?
    (b) What statement can you make about the number of beans in a relative mass of each type of bean?
    (c) What statement can you make about the mass of a bunch of beans whose relative mass is: 3.5? 2.6? 17?
2. (a) Calculate the number of bunches in 1000 beans.
    (b) If you wanted 1000 of the heaviest bean used in this experiment, how could you measure out that approximate number without counting?

3. Calculate the number of beans in:
    (a)   15 bunches of garbanzo beans.
    (b)   3.0 g of a bean whose relative mass is 1.2.
4. What would be the mass of 35 bunches of the heaviest bean?

DISCUSSION II: WAYS TO COUNT ATOMS AND MOLECULES

Knowing the relative mass of a bean and the number of beans in a bunch, we have seen that we could use the bunch as a unit with which to measure out given amounts or numbers of beans. Chemists use a remarkably similar system for measuring amounts of atoms and molecules.

By methods which need not concern us here, it is possible to determine the relative masses of atoms and molecules. Roughly speaking, we can say that we base the relative masses on the average mass of the smallest atom, hydrogen, in the same way that we based the relative mass of a bean on the lightest bean. Using that system, we obtain the relative masses of the atoms and molecules listed below:

PARTICLE	RELATIVE MASS	PARTICLE	RELATIVE MASS
H atom	1.0	O atom	16.0
He atom	4.0	H_2O molecule	18.0
C atom	12.0	CH_4 molecule	16.0

Given the relative masses, we can say that a helium atom is four times as heavy as a hydrogen atom, and that a CH_4 molecule is four times as heavy as a helium atom.

In counting atoms and molecules, and ions as well, we use a unit called the "mole." A mole is equal to a certain number of atoms or molecules, in the same way that a bunch is equal to a certain number of beans. We define the size of the mole in the same way as we did the size of the bunch, namely, as that number of atoms or molecules which has a mass in grams equal to their relative mass. A mole of hydrogen atoms weighs 1.0 gram. A mole of carbon atoms weighs 12.0 grams. A mole of water weighs 18.0 grams. The number of atoms or molecules in a mole is very large, since atoms and molecules are very small. That number is called Avogadro's number and is a constant equal to about 6×10^{23}. Clearly, it is much larger than the number of beans in a bunch! Knowing its value, we can use the mole, as we did the bunch, to count particles and to determine their masses.

Relative masses of atoms are not determined in quite the way we have indicated in the previous paragraph. If they were, then the relative mass of H would be 1.0000, rather than the actual value, 1.0079. In the system now used, the relative mass of the carbon-12 isotope is taken to be exactly 12, and the relative masses of the other atoms are based on it. For practical purposes the results are essentially the same as if we take the relative mass of H to be 1. At the beginning it's probably easier to see the principles if you base the system on hydrogen, so that is why we described it that way. Relative masses of atoms are the same as the atomic masses discussed in Chapter 2. You can find values of the atomic masses inside the back cover of the text.

CALCULATIONS AND QUESTIONS: II

The following questions involving atoms, molecules, and moles are very similar to those you answered involving beans and bunches.

1. (a) If a mole is defined as the number of H atoms in 1.0 g of hydrogen, how many H atoms are there in a mole?
   (b) How many He atoms are there in a mole? How many H_2O molecules? How many Na^+ ions? How many $C_{12}H_{22}O_{11}$ molecules?
   (c) How many grams will a mole of C atoms weigh? A mole of H_2O molecules? A mole of Fe atoms?
2. (a) How many moles are there in a sample containing 12×10^{23} atoms of He? 6×10^{20} O atoms? 1000 H_2O molecules?
   (b) If you wanted a sample of H_2O containing 6×10^{23} molecules, how could you measure out that approximate number, without counting?
3. Calculate the number of atoms or molecules in:
   (a) 15 moles of CH_4 molecules
   (b) 3.0 grams of H_2O molecules
   (c) 2.0 moles of Fe atoms
   (d) 4.0 grams of Fe atoms
4. What would be the mass of 20 moles of carbon atoms?
5. Calculate the mass of one mole of the lightest bean used in this experiment. Contrast that mass with the mass of the earth, 6×10^{27} grams.

EXTENSIONS

1. Assuming that the "bunch" is redefined as the number of beans in 100 g of the heaviest bean in this experiment, calculate the new value for the number of beans in a bunch. Calculate the new mass for 1 bunch of the lightest bean used in this experiment. Check the new values by mass measurements.
2. Determine the new value of Avogadro's number if the mole were to be redefined as "the number of atoms in exactly 10 grams of hydrogen." Using this definition, what would be the mass of (a) one mole of carbon atoms and (b) one mole of CH_4 molecules?

FINDING THE SIZE OF A MOLECULE AND A VALUE FOR AVOGADRO'S NUMBER

OBJECTIVES

1. To learn how laboratory measurements on a macroscopic scale can be related to molecular properties.
2. To see the advantages and limitations of models for solving problems.
3. To gain experience in working with large and small numbers.
4. To apply one of the methods that can be used to find molecular sizes and a value for Avogadro's number.

DISCUSSION

In Experiment 4 we saw that by using the mole as a unit, one can conveniently count very small particles like atoms and molecules. The number of particles in a mole is always the same and is called Avogadro's number. This number can be evaluated by several different methods and is found to equal 6.022×10^{23}. In this experiment we will carry out a simple procedure which will allow us to find approximate values for the size of a molecule and Avogadro's number.

Since molecules are very small and Avogadro's number is very large, any experiment designed to find such magnitudes with ordinary equipment must, directly or indirectly, allow the measurement of either a very small or a very large quantity. In this experiment we will attempt to measure a very small magnitude, namely the dimension of a molecule.

In order to determine the size of a molecule, we work with a substance whose molecules have a rather unique property. That property is that, under proper conditions, the substance will spread out on a water surface until it forms a thin film which is only one molecule thick. In this experiment we measure the area of such a film. Since the film behaves as a typical solid, its area is related to its thickness and its volume.

$$\text{volume} = \text{area} \times \text{thickness} \tag{5.1}$$

The volume of the film we can calculate from the mass of substance in the film and the known density of that substance:

$$\text{volume} = \frac{\text{mass}}{\text{density}} \tag{5.2}$$

Then, from the volume and area of the film we determine its thickness, which is also equal to the length of a molecule of the substance forming the film:

$$\text{thickness} = \frac{\text{volume}}{\text{area}} = \text{length of a molecule} \tag{5.3}$$

Having found the length of a molecule, we can find its volume, *if* we make an assumption about the molecular geometry. The simplest assumption is that the molecule is cubic, and that is the assumption we will make; on that basis:

$$\text{volume of a molecule} = (\text{thickness})^3 \tag{5.4}$$

Knowing the volume of the film and the volume of a molecule, we can find the number of molecules in the film:

$$\text{number of molecules in film} = \frac{\text{volume of film}}{\text{volume of a molecule}} \tag{5.5}$$

The number of moles of substance in the film is given by the equation:

$$\text{number of moles} = \frac{\text{number of molecules}}{\text{Avogadro's number}} \tag{5.6}$$

We can also calculate the number of moles from the mass of substance in the film and the known molecular mass:

$$\text{number of moles} = \frac{\text{mass}}{\text{mass of a mole}} \tag{5.7}$$

Given the number of molecules and the number of moles in the film, we finally obtain a value for Avogadro's number:

$$\text{Avogadro's number} = \frac{\text{number of molecules}}{\text{number of moles}} \tag{5.8}$$

The actual experiment consists of putting a drop of a solution of the film-forming substance on a clean water surface which has been lightly coated with powdered sulfur. The solvent evaporates and the substance spreads out on the water surface. When the film forms, it pushes back the sulfur, and this makes the boundaries of the film easy to see. From the dimensions of the film we calculate its area. From the volume of the drop of solution, and the concentration of the substance in grams per liter, we can determine the mass of film-forming substance in the drop, and hence the mass of the film:

$$\text{mass} = \text{volume of drop in cm}^3 \times \frac{\text{grams of substance in soln}}{\text{volume of solution in cm}^3} \qquad (5.9)$$

Since the density of the substance and its molecular mass are known, all the quantities in Equations 5.1 through 5.9 can be determined. The key calculations are made in Equation 5.3, where we find the length of the molecule, and Equation 5.8, where we obtain a value for Avogadro's number.

We have noted that our experiment depends on the fact that some substances, if handled properly, form a film one molecule thick on a water surface. The substance we use is stearic acid. Its molecular structure is shown below:

$$
\begin{array}{c}
\text{H}\ \ \text{H}\ \ \text{H}\ \ \text{H}\ \ \text{H}\ \ \text{H}\ \ \text{H}\ \ \text{H}\ \ \text{H}\ \ \text{H}\ \ \text{H}\ \ \text{H}\ \ \text{H}\ \ \text{H}\ \ \text{H}\ \ \text{H}\ \ \text{H}\\
|\ \ \ |\ \ \ |\ \ \ |\ \ \ |\ \ \ |\ \ \ |\ \ \ |\ \ \ |\ \ \ |\ \ \ |\ \ \ |\ \ \ |\ \ \ |\ \ \ |\ \ \ |\ \ \ |\\
\text{H}-\text{C}-\text{C}-\text{C}-\text{C}-\text{C}-\text{C}-\text{C}-\text{C}-\text{C}-\text{C}-\text{C}-\text{C}-\text{C}-\text{C}-\text{C}-\text{C}-\text{C}\!\!\begin{array}{c}\diagup\!\!\diagup\text{O}\\ \diagdown\text{O}-\text{H}\end{array}\\
|\ \ \ |\ \ \ |\ \ \ |\ \ \ |\ \ \ |\ \ \ |\ \ \ |\ \ \ |\ \ \ |\ \ \ |\ \ \ |\ \ \ |\ \ \ |\ \ \ |\ \ \ |\\
\text{H}\ \ \text{H}\ \ \text{H}\ \ \text{H}\ \ \text{H}\ \ \text{H}\ \ \text{H}\ \ \text{H}\ \ \text{H}\ \ \text{H}\ \ \text{H}\ \ \text{H}\ \ \text{H}\ \ \text{H}\ \ \text{H}\ \ \text{H}
\end{array}
$$

The molecule consists of a long chain of carbon atoms which, except for the last carbon atom, are all attached to hydrogen atoms. This long chain has the structure of a typical wax, and is very insoluble in water. The $-\text{C}{\diagup\!\!\diagup\text{O} \atop \diagdown\text{O}-\text{H}}$ group on the end of the molecule is similar in structure to water, and is very soluble in water. On the water surface the stearic acid molecules are oriented perpendicularly, as in Figure 5–1, with the waxy chain above the surface and the $-\text{C}{\diagup\!\!\diagup\text{O} \atop \diagdown\text{O}-\text{H}}$ in the surface. When the film forms, the stearic acid molecules pack together much like cordwood, with all the chains parallel. There are very few molecules that behave like stearic acid on a water surface, but it is ideal for the purposes of our experiment.

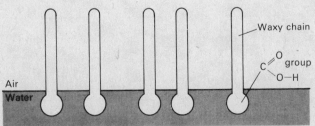

FIGURE 5–1 Stearic acid molecules on a water surface.

PRELIMINARY STUDY

1. Review Section 1.3 in your text on the use of conversion factors.
2. If a circular film with a diameter equal to 10 cm is made up of cubic molecules having an edge equal to 25×10^{-8} cm, how many molecules are there in the film? (Ans. 1.3×10^{15})
3. How much stearic acid is there in a drop of solution having a volume of 0.040 cm^3 if 1000 cm^3 of the solution contains 0.100 gram of stearic acid? (Ans. 4.0×10^{-6} g)

PROCEDURE

1. Obtain a large shallow tray and rinse with tap water to be certain that it is free of any soap or detergent. Fill the tray with water to a depth of about one centimeter.
2. Sprinkle a very small amount of powdered sulfur evenly over the water's surface.
3. Obtain a capillary pipet (about 1 mm in diameter) and carefully place one drop of a stearic acid-cyclohexane solution on the water surface. Measure the diameter of the resulting circular film three ways and find its average diameter.
4. Add a second drop to the center of the film and determine its average diameter in the same manner as before. If space permits repeat the procedure again with a third drop.
5. Determine the volume of one drop from the capillary pipet by counting the drops of cyclohexane required to fill a 10 cm^3 graduated cylinder to the 1 cm^3 mark. Verify your result by counting the number of drops needed to raise the level from the 1 to the 2 cm^3 mark.

SAMPLE DATA TABLE			EXPERIMENT 5
	Film with 1 drop	Film with 2 drops	Film with 3 drops
Diameter 1	_____ cm	_____ cm	_____ cm
Diameter 2	_____ cm	_____ cm	_____ cm
Diameter 3	_____ cm	_____ cm	_____ cm
Avg Diameter	_____ cm	_____ cm	_____ cm
Concentration of stearic acid soln _____ g/1000 cm^3		Calibration of pipet _____ drops/cm^3	

CALCULATIONS AND QUESTIONS

1. (a) Using the average film diameter find the average area for each film. The area of a circle is related to its diameter (d) by the formula:

$$Area = \frac{\pi d^2}{4}$$

   (b) Calculate the area per drop for each film. How do these values compare? What can you conclude about the relative number of molecules on the water's surface for each drop added? Does this support the assumption that the stearic acid is in a monomolecular layer?
   (c) Calculate the average area per drop for the films.
2. Using the data from Procedure 5, calculate the volume of one drop of stearic acid solution.
3. Find the mass of stearic acid in one drop of solution (Eqn. 5.9).
4. Find the volume of stearic acid in one drop of solution (Eqn. 5.2). This is the volume of the film formed by one drop of solution. The density of stearic acid is 0.85 g/cm^3.
5. Using the result from 1(c), calculate the thickness of the film and the length of a stearic acid molecule (Eqn. 5.3).
6. Assuming stearic acid molecules are cubic, find the volume of a molecule (Eqn. 5.4).
7. Using the result of 4, calculate the number of molecules in the film formed by one drop of solution (Eqn. 5.5). (It's a big number.)
8. From the result of 3 and the fact that the molecular mass of stearic acid is 284, calculate the number of moles of acid in the film formed from one drop of solution (Eqn. 5.7).
9. Find a value for Avogadro's number from your data (Eqn. 5.8). How does it compare with the known value? Suggest some reasons for the discrepancy.

EXTENSIONS

1. In calculating Avogadro's number we assumed that the stearic acid molecule is cubic. It is clear from the structure of the molecule that such is not the case. The length (l) of the molecule is greater than its width. If you make a model of the molecule, it turns out that the width (w) is about 1/5 of the length. Calculate the volume of a stearic acid molecule on the basis of this rather better approximation of its geometry:

$$V = lw^2$$

2. On the basis of the new molecular volume, find the number of molecules in the film formed from one drop of solution. From this, calculate a new value for Avogadro's number. Does comparison of this value with the one you found earlier appear to indicate that the stearic acid molecule is longer than it is wide? Give your reasoning.

3. One of the more accurate methods for finding Avogadro's number is based on data obtained from studies of crystals with x-rays. The structure of the NaCl crystal has been determined by such studies to be as in the figure opening Chapter 10 in the text. NaCl has the following properties:

    Molar mass: 58.45 g   Density of solid: 2.165 g/cm³

    Distance between Na^+ and Cl^- ions in the crystal: 2.814×10^{-8} cm

    Find Avogadro's number from this information. Hint: Calculate the volume of 1 mole of NaCl. Taking that volume to be a cube, find the length of the cube edge. How many ions would fit on this edge, given the distance between ions? How many ions would fit in the cube containing a mole of NaCl? How does this number relate to Avogadro's number?

DETERMINING THE SIMPLEST FORMULA OF A COMPOUND

OBJECTIVES

1. To make a compound from its elements.
2. To determine the simplest formula of a compound.
3. To describe a chemical reaction in terms of an equation.

DISCUSSION

In a compound the atoms of different elements are present in numbers whose ratio is usually an integer or a simple fraction. The simplest formula of the compound expresses that ratio and also, if the compound is nonmolecular, serves to identify it. For example, the simplest formula for potassium chlorate, $KClO_3$, tells us that in that compound, for every K atom there is a Cl atom and three O atoms. The K:Cl atom ratio is 1:1, and the K:O atom ratio is 1:3. An important job of the chemist is to find the simplest formulas of any new compounds he or she discovers.

Simplest formulas are determined by establishing the mass of each element present in a given mass of the compound. From those masses one finds the number of moles of each element. The mole ratio must equal the atom ratio in the compound and from that ratio the simplest formula is easily found. A typical calculation is given in Example 3.4 in the text.

To find the mass of each element in a compound one must carry out at least one chemical reaction involving that compound and other known substances. Sometimes it is possible to form the compound directly from the elements. This is called a synthesis reaction. In this experiment we will form, or synthesize, magnesium oxide by burning magnesium in the oxygen in the air:

$$\text{magnesium} + \text{oxygen} = \text{magnesium oxide}$$

By weighing the magnesium before the reaction occurs, and the magnesium oxide produced by the reaction, one can calculate the mass of oxygen that reacted with the magnesium. To do this we need to use the Law of Conservation of Mass. From the masses of magnesium and oxygen in the magnesium oxide we find the simplest formula of that compound. To obtain good results in this experiment you must perform each step carefully, and must be particularly careful to make weighings as accurately as possible.

PRELIMINARY STUDY

1. Review the method of calculating simplest formulas in Section 3.1 of the text.
2. Review the procedure for writing chemical equations in Section 3.2.
3. Practice Problem: In an experiment, 0.87 g of silver reacted completely with sulfur and formed 1.00 g of silver sulfide. Find the simplest formula of the product. (Ans. Ag_2S)
4. When aluminum burns in oxygen, aluminum oxide is formed. The chemical formulas of aluminum, oxygen, and aluminum oxide are Al, O_2, and Al_2O_3. Write the balanced equation for the reaction of aluminum with oxygen.

PROCEDURE

Wear your safety goggles, apron, and gloves during this experiment.

1. Place a clean crucible and cover on a clay triangle on an iron support ring (Figure 6–1). The crucible cover should be tilted on the top of the crucible leaving a small opening. Heat the crucible strongly for about 3 minutes to drive off any moisture. Allow the crucible and cover to cool on a heat resistant pad and then carefully weigh them together.

   CAUTION: Crucibles and covers should be handled with crucible tongs and not with the hands. Hot crucibles should never be placed on a lab bench top; they should be left on the clay triangle or placed on a heat resistant pad.

2. While the crucible is cooling, obtain a piece of magnesium ribbon approximately 50 cm long. Rub the ribbon with a cloth or paper towel to remove any loose surface coating. Coil the ribbon around a pencil. After the cooled crucible has been weighed, add the magnesium and weigh the crucible, lid, and magnesium.
3. With the lid off, heat the crucible. Increase the temperature gradually. When the magnesium ribbon glows red, or ignites, cover the crucible quickly and reduce the amount of heat applied. (If the magnesium actually ignites and burns, the oxide will tend to be vaporized and driven from the crucible. To prevent any loss, the crucible must be covered immediately if you observe an orange or white flame.) After about a minute, carefully remove the cover (to let in more oxygen) and heat until you observe that the magnesium is

glowing, or ignites. At that point replace the cover quickly and reduce the heat. Continue this procedure until no further reaction occurs on heating. Then cover the crucible, leaving a small opening, and heat strongly for a few minutes. Let the crucible cool.

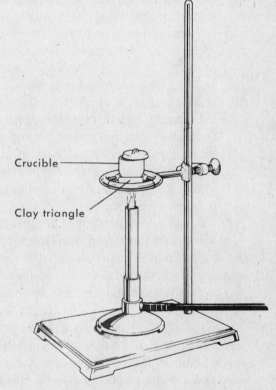

FIGURE 6-1 Heating a crucible. When it is necessary to admit air or to drive out a vapor, the crucible cover is kept off-center to leave a small opening.

Crucible

Clay triangle

4. When the crucible is cool, remove the cover and put it on the bench top. Use a stirring rod to grind the contents of the crucible into small particles. Rinse the particles remaining on the stirring rod into the crucible with about ten drops of distilled water. Replace the cover, leaving a small opening. Heat gently until the water begins to boil. Remove the burner, and sniff the vapor to see if it has any odor. Then, continue heating until all the water is boiled off and the residue is thoroughly dry (about five minutes). Let the crucible cool, and then carefully weigh the crucible, its lid, and the magnesium oxide.
5. Clean the crucible with water as best you can. It is possible that the reaction will damage the glaze on the crucible, so it may not be easy to obtain a smooth inside surface.

SAMPLE DATA TABLE	EXPERIMENT 6
mass of crucible + lid _____ g	
mass of crucible + lid + Mg _____ g	
mass of crucible + lid + product _____ g	

CALCULATIONS AND QUESTIONS

1. (a) Calculate the mass of magnesium used in the reaction and the mass of magnesium oxide produced.
    (b) Calculate the mass of oxygen that reacted with the magnesium sample.
    (c) How many moles of magnesium atoms are there in the magnesium oxide? How many moles of oxygen atoms are there in the oxide?
    (d) What is the Mg/O mole ratio? What is the Mg/O atom ratio in magnesium oxide?
    (e) In most simple compounds the atom ratio can be expressed as a ratio of small integers, such as 1:1, 1:2, 2:1, 3:2, and so on. Round off your experimental atom ratio to obtain a ratio like one of those cited. Given that ratio, what is the simplest formula of magnesium oxide?
2. (a) Using the masses of magnesium and magnesium oxide obtained experimentally, calculate the % Mg in magnesium oxide.
    (b) Using the simplest formula obtained above (le) calculate the % Mg in magnesium oxide.
    (c) Calculate your percent experimental (accuracy) error, assuming the true value is that obtained from the formula.
3. Write the balanced equation for the reaction of magnesium with oxygen gas to form magnesium oxide. The chemical formula for magnesium is Mg. Since oxygen gas is molecular, we use its molecular formula, O_2, in equations. Use (s) or (g) after the formulas to denote that a substance is a solid or a gas.
4. Air is a mixture of nitrogen and oxygen gases. When magnesium burns in air, a small amount of solid magnesium nitride, Mg_3N_2, is formed along with the oxide. Noting that nitrogen gas has the molecular formula N_2, write the equation for the reaction between magnesium and nitrogen gas to form magnesium nitride.
5. If water is added to magnesium nitride and the mixture is heated, the nitride is converted to the oxide. This was done in Procedure 4 to make sure we got just magnesium oxide as a product. Ammonia gas, NH_3, is also produced in the reaction, and you may have detected its odor when you heated the mixture to boiling. Write the balanced equation for the reaction between magnesium nitride and water.

EXTENSIONS

1. Obtain class data for the experimental values of the percent of Mg in the products. Calculate the average % Mg for the class and the class experimental error.
2. Explain the effect on your answer for % Mg (high, low, or unchanged) if the following errors had been made:
    (a) a small amount of Mg remained unreacted.
    (b) some of the product was spilled after heating but before weighing.
    (c) all of the water added in Procedure 4 was not boiled off.
    (d) the product contained some Mg_3N_2.

OBSERVING CHEMICAL REACTIONS AND WRITING EQUATIONS

OBJECTIVES

1. To gain experience in observing the changes which take place during chemical reactions.
2. To learn to write chemical equations to describe chemical reactions.
3. To become more familiar with the properties of some common substances.

DISCUSSION

In the course of chemical reactions changes occur which transform one set of substances (the reactants) into another set (the products). These changes may occur naturally, without any activity on our part, as is the case with the rusting of iron and the rotting of a dead tree. Or humans may make them occur, as is the case when iron ore is converted to iron in a blast furnace, or gasoline is made from coal and water. Although the reactants and products of any reaction contain the same numbers and kinds of atoms, the properties of the reactants and products will usually differ greatly.

In this experiment we will carry out a series of reactions involving copper and its compounds. A few of the reactions may involve physical changes, such as boiling and melting, in which the state but not the nature of substances changes. Reactions involving chemical changes will produce new and different substances. Most of the chemical changes which occur here are easily recognized; a color change takes place, a solid is formed when solutions are mixed, or a visible or smelly gas is given off. In other cases, the changes are more subtle and special procedures are used to detect the products.

For each chemical reaction which occurs you will be asked to write a balanced chemical equation. The reactants and products will either be apparent or you will be given their nature and formulas. Some of the reactions will occur in water solutions and will involve ions. $CuSO_4$, a typical salt, in solution exists as Cu^{2+}

and $SO_4{}^{2-}$ ions, and not as $CuSO_4$ molecules. In solution, NaOH, a common hydroxide, exists as Na^+ and OH^- ions, and H_2SO_4, a well-known acid, exists as H^+ and $SO_4{}^{2-}$ ions. Equations for reactions involving solutions of these substances are written in terms of the *ions* that react. For example, in one part of the experiment you will mix a solution of $CuSO_4$ with one of NaOH. A precipitate of $Cu(OH)_2$ forms. The reaction which occurs involves the Cu^{2+} ions in the $CuSO_4$ solution and the OH^- ions in the NaOH solution. The equation for the reaction is:

$$Cu^{2+} (aq) + 2\ OH^- (aq) \rightarrow Cu(OH)_2(s)$$

In words, the equation states that one mole of Cu^{2+} ions reacts with two moles of OH^- ions in solution to produce one mole of solid $Cu(OH)_2$. The $SO_4{}^{2-}$ ions and the Na^+ ions in the solution are not affected and so do not appear in the equation. The symbol (aq) indicates that the species is in water solution. It would *not* be correct to write the equation for the reaction as:

$$CuSO_4(aq) + 2\ NaOH(aq) \rightarrow Cu(OH)_2(s) + Na_2SO_4(aq)$$

because that would imply that $CuSO_4$, NaOH, and Na_2SO_4 in solution exist as molecules. These substances are all completely ionized in solution, so we write the equation to clearly indicate that fact.

PRELIMINARY STUDY

1. Review the procedures for writing and balancing chemical equations in Section 3.2 of your text.
2. Practice balancing equations for the following reactions.
   (a) Sodium metal reacts with oxygen gas, O_2, to form sodium oxide, Na_2O.
   (b) Sodium oxide reacts with water to form a solution containing Na^+ and OH^- ions.

PROCEDURE

Wear your safety goggles, apron, and gloves while performing this experiment.

In this experiment we will carry out several different reactions. Your instructor will indicate which parts of the experiment you should perform. For each reaction, record your observations in words; also, note the formulas of all reactants and products. Take account of the fact that all of the substances in solution will exist as ions. If you need to wait for one step in one of the procedures to be completed, you may go on to the next one.

1. (a) Make a loose wad of copper wool and place it in a crucible. Place the crucible, uncovered, on a clay triangle on an iron ring and heat strongly for a few minutes. The black product is copper(II) oxide, CuO. The reactants are copper metal, Cu, and oxygen gas, O_2, from the air.

(b) Using crucible tongs, pour the product into a 50 ml beaker. Add about 5 ml of dilute sulfuric acid, 3M H_2SO_4,* and stir the mixture. Set up a funnel and filter the mixture into a 50 ml beaker. Note the characteristic color of Cu^{2+} ions in the filtrate. Water is also formed in this reaction. (In this reaction, the reactants are a solid, CuO, and the H^+ ion from the solution of H_2SO_4. The sulfate ion, also present in the solution, does not enter into the reaction.)

(c) Add 5 ml dilute sodium hydroxide, 6M, NaOH, to the filtrate while stirring. A precipitate will form. Add additional NaOH while stirring until precipitation appears to be complete. The precipitate is copper(II) hydroxide, $Cu(OH)_2$. (Note that in this reaction the reactants are ions in solution.)

(d) Pour off about 1 ml of the above mixture into a small test tube. Add dilute sulfuric acid drop by drop until the precipitate is dissolved. Identify the colored product. The other product is water. At this point you should be able to correctly write the formulas for the reactants; there are two of them: one is a solid and the other is an ion.

(e) Gently boil the remaining mixture from (c) in the beaker until a reaction takes place. One of the products is copper(II) oxide, CuO. (In this reaction a solid is converted to another solid; water is also formed.)

2. (a) Add about 1 g of solid copper sulfate pentahydrate, $CuSO_4 \cdot 5\ H_2O$, to a small test tube and heat strongly for a few minutes. Note any substance forming on the side of the tube. Identify the products. (The reaction involves dehydration of the blue copper sulfate.)

(b) After the tube has cooled to room temperature, add about 5 drops of water to the product. At the same time note any temperature change by feeling the tube. The reaction which occurs is very simply related to the one in (a). Can you see how? Since there are no solutions, there are no ions in either reaction.

3. Add a clean iron nail to about 5 ml of a copper sulfate, $CuSO_4$, solution. Allow to stand for a few minutes, and note any changes that occur. One of the reaction products is the ferrous ion, Fe^{2+}. (The reaction involves a metal reacting with an ion to produce a different metal and ion.)

4. A solution of calcium hydroxide, $Ca(OH)_2$, sometimes called limewater, can be used to test for the presence of carbon dioxide gas, CO_2. When CO_2 is bubbled through limewater, a fairly complicated reaction occurs to form a precipitate of calcium carbonate, $CaCO_3$:

$$Ca^{2+}(aq) + 2\ OH^-(aq) + CO_2(g) \rightarrow CaCO_3(s) + H_2O$$

(a) Obtain a right-angle tube fitted with a stopper from your instructor. Add limewater to a small test tube until it is half-full. Then put about 1 gram of copper carbonate, $CuCO_3$, in another test tube and clamp the tube to a ring stand as shown in Figure 7–1. Add 5 ml dilute sulfuric acid to the solid. Quickly stopper the clamped tube and let the gas being

---

*The 3M before H_2SO_4 means that the solution was made by dissolving 3 moles of H_2SO_4 per liter of solution. 6M NaOH would be made with 6 moles NaOH per liter of solution (see Section 13.1 in the text).

evolved bubble through the limewater until the bubbles have stopped. Examine the limewater and the solution in the clamped tube. Identify two of the products. The third one is water. (There are two reactants, one of which is an ion.)

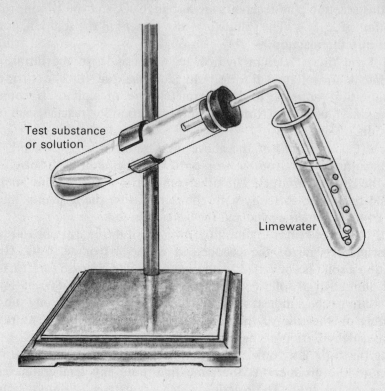

FIGURE 7–1   Test for carbon dioxide. If the sample gives off CO_2, a precipitate will form in the limewater solution.

(b) To a clean dry test tube add about 1 gram solid $CuCO_3$ and clamp the tube as in Figure 7–1. Attach the stoppered right-angle tube. Heat the solid, bubbling any gas that is evolved through fresh limewater as before. Continue heating until no further reaction occurs. Remove the tube of limewater and *then* let the hot tube cool.

CAUTION: Do *not* stop heating before you have taken away the limewater tube, or water will be sucked back into the hot reaction tube and will break it.

Identify the two products of the reaction. There is one reactant.

(c) After the black residue from (b) is cool, add about half its volume of powdered charcoal. Charcoal is essentially carbon, C. Stir the mixture with a stirring rod. Attach the stoppered right-angle tube. Heat the solid mixture gently and then strongly, testing any evolved gases with fresh limewater. Once started, the reaction tends to occur rapidly. Remove the limewater before gas evolution has stopped. Let the test tube cool and examine the residue. There are two products and two reactants.

SAMPLE DATA TABLE			EXPERIMENT 7
Procedure	Reactants	Products	Observations
1(a)			
1(b)			
etc.			

CALCULATIONS AND QUESTIONS

1. In each part of each procedure a chemical reaction involving copper or one of its compounds occurred. Using the reactants and products you have recorded, write a balanced chemical equation for each reaction. Label each reaction by its procedure number. To identify the state of each species, use the symbols (s), (l), (g), or (aq) to denote that the species was a solid, liquid, gas, or an ion in water solution. You may assume that any substance in solution in the experiment exists as ions.

2. (a) What chemical property is shared by copper oxide, copper hydroxide, and copper carbonate?

    (b) What evidence do you have that the blue color of a copper sulfate solution is due to the Cu^{2+} ion?

    (c) What chemical test could you use to distinguish between copper sulfate pentahydrate and copper carbonate, both of which are blue solids?

EXTENSION

1. Test the solubility of copper carbonate by adding a pinch of it to about 5 ml of water. Predict what will happen if solutions of copper sulfate and sodium carbonate, Na_2CO_3, are mixed. Test your prediction and write a balanced equation for any reaction which occurs.

THE SPECIFIC HEAT OF LIQUIDS AND SOLIDS

OBJECTIVES

1. To determine the specific heats of a solid and a liquid.
2. To learn the techniques of calorimetry.
3. To better understand the concepts of heat, temperature, and specific heat.

DISCUSSION

Heat is a form of energy which passes from an object of high temperature to an object of lower temperature when the two objects are in contact. When heat flows into a substance the temperature of that substance will increase. The quantity of heat, Q, required to cause a temperature change, Δt, is proportional to the temperature change and to the mass, m, of the substance being heated. These relationships may be expressed as

$$Q = C \times m \times \Delta t \qquad (8.1)$$

The proportionality constant, C, is called the specific heat of the substance.

When the mass is expressed in grams, the specific heat can be considered as the amount of heat required to raise the temperature of one gram of the substance by one degree Celsius. The calorie, a heat unit, is defined as the amount of heat required to raise the temperature of one gram of water by one degree Celsius. For this reason, the specific heat of water is 1.00 cal/g°C. In Part I of this experiment the specific heat of an unknown liquid will be determined from that of water by comparing the temperature change of equal masses of water and the liquid when the same amount of heat is added to each. Any difference in the amount of temperature change is due to a difference between the specific heat of water and that of the liquid.

Heat flow is often measured in a device called a calorimeter. A calorimeter is a container made with insulating walls so that essentially no heat is exchanged between its contents and the surroundings. In Part II of our experiment we will use a pair of nested styrofoam coffee cups as our calorimeter. A chemical or physical change can take place inside the calorimeter in such a way that any heat effects involve only the contents of the calorimeter. In this part of the experiment, a known mass of a solid is heated to a known temperature and poured into a calorimeter containing a known mass of water at a known temperature. Heat will flow from the solid to the water until the temperature of the solid is equal to that of the liquid. In this process, the amount of heat lost by the solid is equal to the heat gained by the water (assuming no heat losses).

Amount of heat leaving the solid = Amount of heat absorbed by the water

By Equation 8.1,

$$C_s \times m_s \times \left| \Delta t_s \right| = C_{H_2O} \times m_{H_2O} \times \left| \Delta t_{H_2O} \right| \qquad (8.2)$$

From the masses of the solid and the water and the changes in their temperatures, you can calculate the specific heat, C_s, of the solid.

PRELIMINARY STUDY

1. Explain the difference between heat, temperature, and specific heat. What units are used to express these properties?
2. Practice Problem: Determine the number of calories absorbed when 200 cm³ of H_2O is heated from 25.2°C to 37.7°C. (Ans. 2500 cal)

PROCEDURE

Wear your safety goggles, apron, and gloves while performing this experiment.

Part I: Specific Heat of a Liquid

1. Obtain an electrical hot plate. Turn it on and allow it time to reach a constant temperature. Set the thermostat at a medium setting.
2. (a) Obtain two clean, dry 150 cm³ beakers whose masses are within 2 g of each other. Measure out precisely 100 ml (100 g) of distilled water in a graduated cylinder and pour it into one of the beakers. Label the beaker.
   (b) Obtain an unknown liquid and record its number and density. Calculate the volume of 100 g of the liquid. Measure out carefully the calculated volume of the unknown liquid and pour it into the second beaker. Label the beaker.
3. (a) Record the temperatures of both liquids and then place both beakers on the hot plate simultaneously. Put a watch glass on top of each beaker to minimize heat loss.
   (b) Allow the liquids to heat evenly by periodically moving the beakers to different positions without lifting them from the hot plate. Using separate stirring rods, stir each liquid periodically to distribute the heat. After

about 5 minutes of heating (or as soon as the temperature of the water has increased about 30°C) simultaneously remove both beakers from the hot plate. Immediately stir each liquid and record its maximum temperature.

Part II: Specific Heat of a Solid

1. (a) Fill a 400 ml beaker about 3/4 full of water and cover it with a watch glass. Put it on the hot plate and heat it to the boiling point. Set the thermostat at the highest setting. Go on to the next step while waiting for the water to boil.

   (b) Obtain from your instructor a sample of an unknown metal in a large, dry, stoppered test tube. Weigh the test tube and metal. Pour the metal into a dry beaker and weigh the empty tube. Put the metal back in the tube and replace the stopper loosely.

2. (a) Place the tube containing the metal in the boiling water bath. The beaker should contain enough water so that all of the metal is below the water's surface. Allow the metal to heat at least five minutes to ensure that it reaches the temperature of the water. Measure and record the temperature of the boiling water.

   (b) While the metal is heating assemble the calorimeter by placing two nested styrofoam coffee cups in a 250 ml beaker (Fig. 8–1). Contact between the cups and the beaker should occur only near the top of the outer cup. Measure out carefully 50 ml (50 g) of distilled water in a graduated cylinder and pour the water into the calorimeter. Measure and record the water temperature.

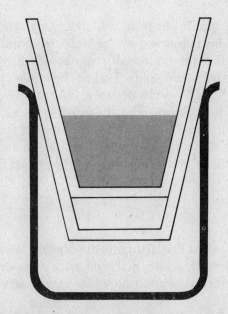

FIGURE 8-1 Coffee cup calorimeter.

(c) When the metal is up to temperature, remove the tube from the boiling water, remove the stopper and quickly pour the metal into the water in the calorimeter. Be sure that no hot water from the test tube gets into the calorimeter water and that no water splashes out of the calorimeter when you pour in the metal. Stir the water in the calorimeter and record its maximum temperature.

SAMPLE DATA TABLE   **EXPERIMENT 8**

Part I: unknown liquid no. _____

density of unknown _____ g/cm^3 vol. of 100 g of unknown _____ cm^3

temp of H_2O (initial) _____ $°C$ temp of unknown (initial) _____ $°C$

temp of H_2O (final) _____ $°C$ temp of unknown (final) _____ $°C$

Δt of H_2O _____ $°C$ Δt of unknown liquid _____ $°C$

Part II: unknown metal no. _____

mass of metal and test tube _____ g   mass of test tube _____ g

mass of H_2O (calorimeter) _____ g   temp of H_2O (initial) _____ $°C$

temp of H_2O (bath) _____ $°C$ temp of H_2O (final) _____ $°C$

Δt of metal _____ $°C$ Δt of H_2O _____ $°C$

CALCULATIONS AND QUESTIONS

1. Using the specific heat of water and its temperature change calculate the amount of heat absorbed by the water from the hot plate in Part I (Eqn. 8.1).
2. Assuming that the unknown liquid absorbed the same amount of heat as the water, calculate the specific heat of the liquid (Eqn. 8.1).
3. Using the specific heat of water and its temperature change calculate the amount of heat absorbed by the water in the calorimeter in Part II.
4. Assuming that the unknown metal is heated to the temperature of the boiling water bath and that all the heat lost by the metal goes into the calorimeter water, calculate the specific heat of the metal (Eqn. 8.2).

EXTENSIONS

1. It is assumed in Part II of the experiment that not only is the calorimeter a good insulator but also that it absorbs essentially no heat from its contents. Although these are good approximations, some heat will actually be lost to

the surroundings, and some will be absorbed by the calorimeter walls and the thermometer. Explain whether these effects would result in specific heat values that are larger or smaller than the true values.

2. The specific heat of many metals is related in a simple way to the mass of a mole of the metal. The relationship, known as the Law of Dulong and Petit, states that about 6 calories are required to raise the temperature of one mole of a metal by one degree centigrade.

$$C_s \times AM \text{ in g} = 6.0 \text{ cal/}^\circ C$$

Test this relationship by calculating the atomic mass of the metal unknown in Part II from your value of the specific heat. See if you can identify the metal. Check with your instructor to see if you are right. Predict the specific heat of the metal you actually had, using the Law of Dulong and Petit, and calculate your percent experimental error.

HEAT EFFECTS OF CHEMICAL REACTIONS

OBJECTIVES

1. To investigate the heat effects which occur during chemical reactions.
2. To determine the change in heat content, ΔH, for two reactions.

DISCUSSION

Almost all chemical reactions are accompanied by an energy change. The energy change is usually made evident by a heat effect. If heat is evolved the reaction is exothermic; if heat is absorbed the reaction is endothermic. We can tell whether a reaction is exothermic or endothermic by noting its effect on the temperature of the products. When methane gas is burned, the products, carbon dioxide and water, are considerably hotter than the original reactants. As the products cool to the original temperature, heat must go out of (exit from) the products and so the reaction is *exo*thermic. When they get back to the original temperature the products have a lower heat content (H) than the reactants by the amount of heat which they lost on cooling. If the products of a reaction are cooler than the reactants, heat must go into the products to restore the original temperature, and the reaction is *endo*thermic. The change in heat content is equal to the heat effect for the reaction and may be written as ΔH:

$$\Delta H = H_{products} - H_{reactants} = \text{heat effect}$$

A water calorimeter was used in Experiment 8 to measure the heat transferred from a hot object (the metal) to a cooler one (the water). The same calorimeter technique can be used to measure ΔH for a chemical reaction. In the simplest case, the reaction is made to take place in the water. In this case the heat involved in the reaction is transferred directly to the water in the calorimeter. The heat effect in the water can then be calculated knowing its specific heat, C, the amount of water used, m, and its temperature change, Δt.

$$Q_{H_2O} = C_{H_2O} \times m_{H_2O} \times \Delta t_{H_2O} \tag{9.1}$$

If it is assumed that no heat is lost to the surroundings, then the heat effect in the water will be equal to minus the change in heat content in the reaction:

$$Q_{H_2O} = -\Delta H_{reaction} \tag{9.2}$$

Because one system (calorimeter or reaction mixture) is absorbing heat while the other system is releasing heat, the heat effect terms will have opposite signs. The sum of the two heat effects must be zero, since energy (heat) is conserved in the calorimeter. As the heat effect is directly related to the amount of reactants used, it is customary to express the change in heat content (ΔH) on a per gram or a per mole basis.

PRELIMINARY STUDY

1. Review the writing of thermochemical equations in Chapter 4.
2. Practice Problem: A water calorimeter absorbed 17.8 kcal from the burning of 2.50 g of ethyl alcohol (MM = 46.0). Calculate the value of ΔH for the burning of one mole of ethyl alcohol. (Ans. $\Delta H = -327$ kcal per mole)

PROCEDURE

Part I: Heat of Combustion

1. Light a candle and stick it to a glass plate with hot wax. Extinguish the candle and weigh the candle and plate.
2. Obtain a small can which has two holes punched opposite each other near its top. Using a graduated cylinder, measure out precisely 200 ml (200 g) of tap water and pour it into the can.
3. Obtain a large can from which the ends have been removed and which has holes punched near its base. This can will serve as a wind guard and chimney for the candle. Assemble the candle and two cans as shown in Figure 9–1, using a stirring rod to support the small can on the ring. The small can should be inside the larger one as far as possible and still be about one inch above the top of the candle wick. (You will find the best position most easily if you make the adjustment first without the large can in position.)
4. Lift the small can out of the ring and put it on the lab bench. Read and record the water temperature. Light the candle with a match held with crucible tongs, and immediately reposition the small can.
5. Stir the water periodically and note its temperature. Allow the candle to burn for at least five minutes or until the water temperature has increased 30°C. (Begin Part II while waiting.)
6. After the water is at about 55°C, carefully blow out the candle flame. Continue to stir the water and record the highest temperature reached.
7. Weigh the candle and glass plate, being sure to include all wax drippings.

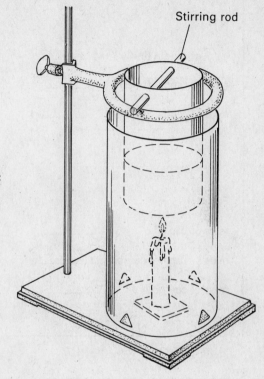

FIGURE 9-1 Apparatus for determining the heat of combustion of a candle.

Part II: Heat of Solution

1. Assemble a calorimeter like the one in Experiment 8, from two nested styro-
    foam coffee cups and a 250 ml beaker. Measure precisely 50 ml (50 g) of
    distilled water in a graduated cylinder and pour it into the calorimeter.
2. Weigh out about 5 g of the compound assigned by your instructor on paper
    which you have previously weighed. Record the masses to the nearest 0.01
    g. Record the formula of the compound used.
3. Measure the temperature of the water to the nearest 0.1°C. Add the com-
    pound to the water and stir so that it will dissolve more rapidly. Record the
    maximum or minimum temperature reached during the solution process.

SAMPLE DATA TABLE   EXPERIMENT 9

Part I

mass of candle (initial)   _____ g   water temperature (initial) _____ °C

mass of candle (final)   _____ g   water temperature (final)   _____ °C

mass of candle (burned)   _____ g   temperature change, Δt_{H_2O} _____ °C

                                                    mass of water   _____ g

Part II   formula of compound   _____

mass of paper   _____ g   water temperature (initial) _____ °C

mass of paper + compound _____ g   water temperature (final)   _____ °C

mass of compound   _____ g   temperature change, Δt_{H_2O} _____ °C

                                                    mass of water   _____ g

CALCULATIONS AND QUESTIONS

1. (a) Calculate the amount of heat absorbed by the water during the combustion of the candle in Part I (Eqn. 9.1).
    (b) Is the combustion reaction exothermic? Why?
    (c) How much heat was absorbed by the water per gram of candle burned?
    (d) How much candle would have to burn to bring a liter of water starting at room temperature, 25°C, to the boiling point?
    (e) The heat of combustion, ΔH, of a candle is about −10 kilocalories per gram. What fraction of the heat given off by the candle was absorbed by the water in the can?
    (f) Assuming that the candle wax has the formula $C_{20}H_{42}$, write the equation for the combustion of the candle.
2. (a) Determine the amount of heat gained or lost by the water, Q_{H_2O}, in the dissolving of the compound in Part II (Eqn. 9.1). Note the sign of Δt.
    (b) Is the solution reaction exothermic? Why?
    (c) What is the sign of ΔH for the solution reaction (Eqn. 9.2)?
    (d) Calculate the number of moles of the compound dissolved.
    (e) Determine the heat of solution in calories per mole by dividing the value of ΔH by the number of moles of compound dissolved.

EXTENSIONS

1. In the experiment on the combustion of a candle, much of the heat given off is not absorbed by the water. Where does the rest of the heat go? Suggest several ways by which the experiment might be modified to help make it possible to actually measure the heat of combustion of one gram of candle.
2. Instant hot packs or cold packs are often used by coaches and trainers to treat injuries when hot packs or ice are not available. The portable packs are usually plastic bags which contain a dry chemical to which water is added. Obtain either kind of pack and determine the heat of solution of one gram of the solid. (Part II).

DETERMINING THE MOLECULAR MASS OF A GAS

OBJECTIVES

1. To determine the mass of a mole of a gas experimentally.
2. To gain skill at correcting gas volumes to standard conditions.
3. To learn a technique for gas generation and collection.

DISCUSSION

The relationships between pressure, volume, and temperature for a gas are much simpler than for a liquid or solid. In particular, they do not depend on the nature of the gas and they can be expressed quite accurately by simple mathematical equations. The key equation for gas behavior, applicable to a fixed amount of any gas undergoing a change, is

$$\frac{P_1 V_1}{T_1} = \frac{P_2 V_2}{T_2} \tag{10.1}$$

In the equation, 1 and 2 refer to initial and final states of the gas. If a sample of gas at P_1, V_1, and T_1 experiences a change in state, it is possible to use Equation 10.1 to calculate any one of the three properties in the final state, given the other two.

Equation 10.1 is often used to find the volume of a gas under standard conditions, STP, at $0°C$ and 1 atm, from a set of initial values P_1, V_1, and T_1. Under such conditions, P_2 equals 1 atm, or 760 mm Hg, and T_2 equals 273 K, the temperature on the absolute temperature scale that corresponds to $0°C$. Solving for V_2, the volume at STP, we obtain, using Eqn. 10.1

$$V_2 = \frac{273 \text{ K}}{1 \text{ atm}} \times \frac{P_1 V_1}{T_1} \tag{10.2}$$

where P_1 is expressed in atm and T_1 in K. V_2 will have the same units as V_1, usually liters.

The importance of knowing the volume of a gas at STP is that it allows you to calculate the number of moles of gas in the sample. Any gas at STP has essentially the same molar volume, 22.4 liters. Conversion of volume at STP to number of moles is therefore readily accomplished:

$$\text{moles of gas in sample} = V_2 \text{ (liters at STP)} \times \frac{1 \text{ mole}}{22.4 \text{ liters}} \qquad (10.3)$$

From the number of moles and the mass of the sample it is easy to calculate the mass of one mole of gas:

$$\text{molar mass} = \frac{\text{mass of gas}}{\text{moles of gas}} \qquad (10.4)$$

The molar mass is numerically equal to the molecular mass and is directly related to the molecular formula of the gas.

In this experiment we will measure the mass of a sample of carbon dioxide at a given temperature, pressure, and volume. From these data we can obtain the volume at STP, and from that volume the molar mass of CO_2. This method offers probably the simplest way to experimentally determine the molecular mass of a substance.

The carbon dioxide in the experiment will be made in a gas generator. In the generator we will carry out the following reaction between marble chips ($CaCO_3$) and hydrochloric acid:

$$CaCO_3(s) + 2\ H^+(aq) \rightarrow CO_2(g) + Ca^{2+}(aq) + H_2O \qquad (10.5)$$

(This reaction is typical of carbonates and was used in Experiment 7 with $CuCO_3$.) The carbon dioxide evolved is dried and collected in a flask at room temperature and atmospheric pressure. The mass of CO_2 is found by subtracting the mass of the flask filled with CO_2 from that of the empty flask. We get the mass of the empty flask, not from its mass when completely evacuated, but rather from its mass when filled with air. From the density of air and the volume of the flask we can determine the mass of air in the flask and hence the actual mass the flask would have if it contained no gas at all.

PRELIMINARY STUDY

1. Review the use of the gas laws in Chapter 5.
2. Practice Problem: Determine the molar mass of a gas which weighs 0.60 g and occupies a volume of 0.500 liter at 20°C and 730 mm of Hg. (Ans. 30 g)

PROCEDURE

Wear your safety goggles, apron, and gloves while performing this experiment.

1. Obtain a clean, dry 250 ml Florence or Erlenmeyer flask (Figure 10–1). Stopper the flask with a one-hole stopper which has a glass tube inserted in the hole. The tube should reach to within 1 cm of the bottom of the flask. Cap the top of the glass tube with a bulb from a medicine dropper. Weigh the flask and the air it contains along with the stopper assembly. From this point on, hold the flask by its rim to minimize the effect of handling on its mass.

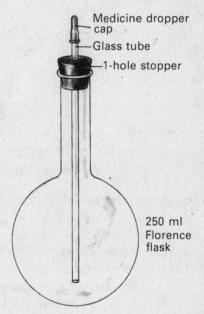

FIGURE 10-1 Flask assembly for measuring the molecular mass of a gas.

2. Put about 20 grams of calcium carbonate (marble chips) in a wide mouth bottle. Then add just enough water to cover the chips. Stopper the bottle with a two-hole stopper which has a right-angle glass tube and a long-necked funnel (or thistle tube) inserted. The arm of the right-angle tube should extend to just below the bottom of the stopper. The funnel tube should reach below the water surface. (If you have to insert the tubes into the stopper yourself, ask your teacher to show you the proper procedure.) Clamp a drying tube to a ring stand at about the height of the arm of the bent tube. (The drying tube contains a drying agent, such as calcium chloride, which absorbs any moisture carried by the gas.) Fit a one-hole stopper, with a short glass tube inserted, to the drying tube. Use pieces of rubber tubing to connect the wide-mouth bottle and the flask to the drying tube as shown in Figure 10–2.

3. Before proceeding, check the entire set-up and have it approved by your instructor. All connections should be tight *except* the stopper on the gas collecting flask, which should be loose. Why? When you are ready, pour some dilute (6 M) hydrochloric acid, HCl, into the funnel from a small beaker. Add the acid a little at a time in order to maintain a steady generation of gas.

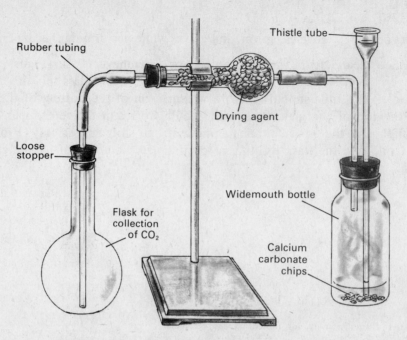

FIGURE 10–2   Apparatus for the generation and collection of CO$_2$.

Continue the generation of gas for about 15–20 minutes. The CO$_2$ evolved is more dense than air so, as the reaction proceeds, carbon dioxide will gradually fill the flask, displacing the air and forcing it into the room.

4. (a)   After 20 minutes, when the flask should be full of CO$_2$, remove the rubber tubing from the collection flask, tighten the stopper assembly, and cap the glass tube. Weigh the stoppered flask of carbon dioxide gas.

   (b)   Mark the level of the bottom of the stopper with a marking pen. Remove the stopper and fill the flask with water to that level. Measure the volume of the water contained in the flask by pouring it into a large graduated cylinder. This value corresponds to the volume of the flask and to the volume of the collected CO$_2$.

5. Record the room temperature and the laboratory pressure as obtained from a barometer. These values are the same as the temperature and pressure of the carbon dioxide.

6. Decant the liquid from the wide-mouth bottle. (Do not dump solids in the sink.) Rinse off the remaining marble chips with water, pour off the water, and discard the chips into a container for solid wastes.

```
SAMPLE DATA TABLE                          EXPERIMENT 10

mass of flask,                      air density        _____ g/cm³
stopper assembly and air  _____ g
                                    temp of CO₂, T₁  _____ °C

mass of flask,                      lab pressure, P₁  _____ mm Hg
stopper assembly and CO₂  _____ g
                                    vol. of flask, V₁  _____ cm³
```

CALCULATIONS AND QUESTIONS

1. (a) Using the density of air at laboratory conditions (provided by your instructor) and the volume of the flask, calculate the mass of air in the flask.
   (b) Using the known mass of air calculate the mass of the "empty" flask and stopper assembly.
   (c) Using the mass of the empty flask and stopper, determine the mass of the collected CO_2.
2. (a) Convert the values of P_1, V_1, and T_1 of the CO_2 in the flask to units suitable for use in the gas law (atm, liters, and K).
   (b) Use the gas law to find the volume that the CO_2 would have under standard conditions (0°C and 1 atm) (Eqn. 10.2).
   (c) Find the number of moles of CO_2 in the sample you collected (Eqn. 10.3).
3. (a) From the mass and number of moles of CO_2 in the sample, calculate the mass of one mole of CO_2 (Eqn. 10.4). This mass is numerically equal to the molecular mass of the gas.
   (b) Calculate your percent experimental error based on the molecular mass of CO_2 as determined from its formula.
4. (a) Using Eqn. 10.5, determine the number of moles of CO_2 which can be produced from 20.0 g of $CaCO_3$.
   (b) Determine the volume that this amount of CO_2 would occupy at 25°C and 1 atm. (Hint: First find the molar volume of CO_2 at 25°C and 1 atm from the value at STP. (Eqn. 10.1 is applicable.)

EXTENSIONS

1. Using the same experimental apparatus as above, determine the molecular mass of oxygen gas. The gas may be produced by the decomposition of hydrogen peroxide, H_2O_2, in the presence of a catalyst. Place about 5 g of manganese dioxide, MnO_2, in the bottom of the bottle. Gradually add 100 ml of a 3% hydrogen peroxide solution. Proceed as in the carbon dioxide experiment.

2. Explain why the set-up used in this experiment would not be satisfactory for determining the molecular mass of hydrogen. Hint: Calculate the approximate mass of hydrogen in the flask at laboratory conditions. How would you modify the experiment to obtain satisfactory results?

VOLUME-MASS RELATIONS IN CHEMICAL REACTIONS

OBJECTIVES

1. To relate the volume of a gas produced to the amount of a reactant consumed.
2. To gain experience in the use of Dalton's Law.
3. To balance a chemical equation using experimental data.

DISCUSSION

Dalton showed that a mixture of gases exerts a pressure which is equal to the sum of the partial pressures of the individual gases. The partial pressure of a gas A is the pressure that gas A would exert if it were in the container by itself. For a mixture of gases A, B, and C:

$$P_{total} = P_A + P_B + P_C \qquad (11.1)$$

When a gas is collected over water, as in this experiment, it becomes saturated with water vapor. The collected gas is a mixture of a "dry gas" and water vapor. According to Dalton's Law the total pressure exerted by the collected gas is:

$$P_{total} = P_{dry\ gas} + P_{H_2O} \qquad (11.2)$$

In a gas collected this way the partial pressure of water vapor, P_{H_2O}, will be constant at any given temperature and equal to the vapor pressure of water at that temperature. In order to use the gas laws to determine the amount of dry gas in the sample it is necessary to know the partial pressure of the dry gas. This can be obtained by subtracting the water vapor pressure, P_{H_2O}, from the total pressure:

$$P_{dry\ gas} = P_{total} - P_{H_2O} \qquad (11.3)$$

The vapor pressure of water at the temperature of this experiment can be found in tables such as Table 11–1.

TABLE 11–1   Vapor Pressure of Water

TEMPERATURE (°C)	PRESSURE (mm Hg)	TEMPERATURE (°C)	PRESSURE (mm Hg)
15	12.8	23	21.0
16	13.6	24	22.4
17	14.5	25	23.7
18	15.5	26	25.2
19	16.5	27	26.7
20	17.5	28	28.3
21	18.6	29	30.0
22	19.8	30	31.8

In this experiment you will measure the volume of hydrogen gas generated by the reaction of a known amount of magnesium metal with hydrochloric acid, HCl. The hydrogen will be collected over water and so will be saturated with water vapor. Using the partial pressure of the hydrogen, the volume of gas collected, and the temperature, you can determine the number of moles of H_2 in the gas sample. From this and the number of moles of magnesium reacted you can find the hydrogen gas:magnesium mole ratio. Using that ratio you should be able to write the balanced chemical equation for the reaction.

PRELIMINARY STUDY

1. Review the use of Boyle's Law, Charles' Law, and Dalton's Law in Chapter 5.
2. Practice Problem: A student collects oxygen gas over water at 23°C. If the total pressure of the wet gas is 730 mm Hg, what is the partial pressure of the oxygen? (See Table 11–1 for the vapor pressure of water.) Convert this pressure to atmospheres. (Ans. 709 mm Hg, 0.933 atmosphere)

PROCEDURE

Wear your safety goggles, apron, and gloves while performing this experiment.

1. Preparation of Metal Sample
   (a) Obtain a strip of magnesium ribbon about 4–5 cm long. Measure and record its length to 0.1 millimeter. Record the mass of one meter of the ribbon as provided by your teacher.
   (b) Wrap a piece of fine copper wire approximately 20 cm long around the magnesium in such a way as to hold the ribbon tightly. (Figure 11–1 [a]) (The copper will not react with the acid.) Leave about 15 cm of wire unspiralled to serve as a handle.

FIGURE 11-1 (a) Copper spiral around Mg; (b) copper spiral installed in buret; (c) spiral and buret during reaction.

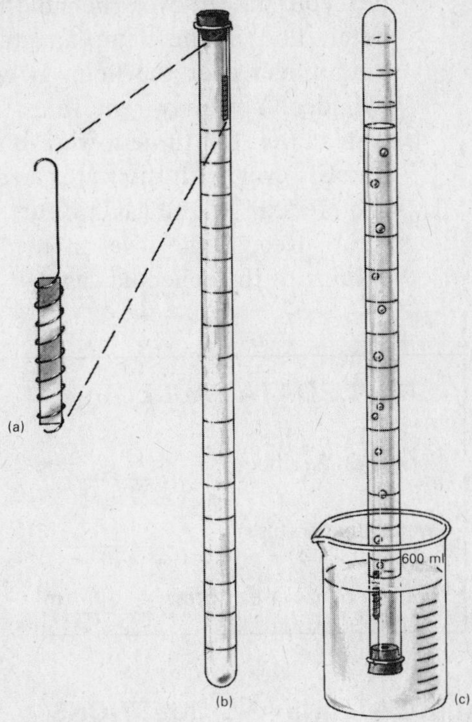

2. Generation of Hydrogen Gas
   (a) Obtain a gas buret and clamp it to a ring stand with its open end up. The buret will serve as the container for the gas. Pour 10 ml of dilute hydrochloric acid (6M HCl) into the buret. Then tilt the buret slightly and add water until the buret is full. Try to avoid mixing the relatively dense acid with the water.
   (b) Fill a 600 ml beaker about 3/4 full of water at room temperature, and place it near the ring stand.
   (c) Slide the magnesium in its copper holder into the buret, hooking the handle over the rim. Stopper the buret tightly with a one-hole stopper; this will keep the Mg sample in place. (Fig. 11–1 [b])
   (d) Put your finger over the stopper hole and invert the buret. Quickly immerse the end of the buret in the beaker. Remove your finger when the stopper is below the water level. Clamp the buret in an upright position, as in Figure 11–1 (c). When the relatively dense hydrochloric acid comes down the tube and reaches the magnesium, a reaction will occur, producing hydrogen gas bubbles.
3. Measurement of the Volume of Gas at Barometric Pressure
   (a) After all of the ribbon has reacted, allow another five minutes for the gas to reach temperature equilibrium. Measure and record the water temperature and the barometric pressure.
   (b) The gas produced in the reaction is not at barometric pressure, since the level of the liquid in the buret and that in the beaker are not equal. To obtain the volume at the pressure of the room we proceed as follows. Fill

a large graduated cylinder with water. Then reach into the beaker and put your finger over the hole in the stopper in the buret so no air can enter. Loosen the clamp and raise the buret out of the beaker, keeping your finger over the hole. Lower the buret into the water in the large cylinder. You may remove your finger when the stopper is below the water level. Continue lowering the buret until the gas-liquid level in the buret is even with the water level in the graduate. Under these conditions the pressure of the gas mixture will equal the barometric pressure in the room. Record the level in the buret to the nearest 0.1 ml. This is the volume of the collected gas.

SAMPLE DATA TABLE   **EXPERIMENT 11**

length of Mg ribbon _____ mm   mass of Mg/meter _____ g/m

water temperature _____ °C   barometric pressure ____ mm Hg

volume of gas at bar. press. _____ ml

CALCULATIONS AND QUESTIONS

1. The gas in the tube consists of a mixture of hydrogen and water vapor. When the liquid levels are equalized the pressure exerted by both gases is equal to the barometric pressure:

$$P_{H_2} + P_{H_2O} = P_{barometric}$$

Obtain the pressure of the water vapor from Table 11–1 at the temperature of the water in the beaker. Using the above equation, determine the partial pressure of the dry hydrogen gas. This is the pressure exerted by H_2 in the gas mixture, P_1.

2. (a) To find the number of moles of H_2 produced in the reaction we need to find the volume the dry H_2 would occupy under standard conditions, 0°C and 1 atm, where the molar volume of a gas is known to be 22.4 liters. The approach is like that in Experiment 10; we use the gas law for a fixed amount of gas:

$$V_2 = \left(\frac{T_2}{P_2}\right) \times \left(\frac{P_1 V_1}{T_1}\right)$$

Here, 2 refers to standard conditions and 1 to the conditions of the experiment. P_1 then is P_{H_2} expressed in atm, V_1 is the volume of collected gas expressed in liters, and T_1 is the water temperature in K. Calculate V_2, the volume in liters that the H_2 would occupy at 273 K and 1 atm.

3. (a) Using the molar volume of a gas at STP, calculate the number of moles of H_2 collected in this experiment (Eqn. 10.3).
   (b) From the length of magnesium ribbon reacted, and the mass per meter (1000 mm), find the mass of magnesium used in the reaction.
   (c) Calculate the number of moles of magnesium used in the reaction. AM Mg = 24.31.
   (d) From the results of (a) and (c) determine the hydrogen gas:magnesium mole ratio. Round off the value to the nearest integer or simple fraction.
4. The reaction between magnesium and hydrochloric acid involves only the H^+ ion present in the solution of HCl. The products are H_2 gas and the magnesium ion; the unbalanced equation for the reaction is:

$$\underline{\quad} Mg\ (s) + \underline{\quad} H^+\ (aq) \rightarrow \underline{\quad} H_2\ (g) + \underline{\quad} Mg^{n+}\ (aq)$$

From the mole ratio found in 3(d) you can find the number of moles of H_2 produced from one mole of Mg. Balance the equation by finding the proper number of moles of H^+ ions consumed per mole of Mg, and the charge on the magnesium ion. To find the latter you need to use the fact that the charges on the left and the right must be equal.

EXTENSIONS

1. Find the number of moles of H_2 produced in the experiment by using the Ideal Gas Law (Eqn. 5.18 in the text). Compare your results and the calculations that are required.
2. Repeat the above experiment using aluminum instead of magnesium. Consult with your teacher for the amount of aluminum to be used. This time, use 20 ml of dilute acid so that the reaction will occur reasonably quickly. Perform the calculations as before.

THE PREPARATION AND PROPERTIES OF PURE OXYGEN

OBJECTIVES

1. To prepare and collect some pure oxygen gas.
2. To carry out oxidation reactions in air and in pure oxygen.

DISCUSSION

Oxygen is our most plentiful element. It exists in compound form in water and in many rocks and minerals. Elemental oxygen makes up about one-fifth of the gas in the atmosphere. Under the appropriate conditions most elements will react with oxygen to form oxides. Metals form solid ionic oxides while non-metals form molecular oxides. The reactions with zinc and sulfur, which are typical, are shown below:

$$Zn(s) + 1/2\ O_2(g) \rightarrow ZnO(s)$$

$$S(s) + O_2(g) \rightarrow SO_2(g)$$

Reactions with oxygen are called "combustion" or, more correctly, oxidation reactions. When a substance reacts with oxygen so rapidly that heat and light are produced, the substance is said to be "burning".

Oxygen gas, O_2, may be prepared by decomposing some of the compounds which contain oxygen. In fact, oxygen was discovered when Priestly heated mercuric oxide, HgO, and succeeded in driving off oxygen gas from this compound. In this experiment you will obtain oxygen gas by using heat to decompose potassium chlorate, $KClO_3$:

$$2\ KClO_3(s) \xrightarrow[MnO_2]{heat} 2\ KCl(s) + 3\ O_2(g)$$

The reaction occurs more rapidly in the presence of MnO_2, which serves as a catalyst for decomposition of $KClO_3$. A catalyst increases the rate of a reaction but is not itself consumed in the reaction.

PRELIMINARY STUDY

1. Review Section 6.2 in the text.
2. Practice Problem: (a) Calculate the number of moles of O_2 gas produced by the decomposition of 12.3 g of $KClO_3$. (b) What would be the volume of the gas under standard conditions? (Ans: 0.150 moles, 3.36 liters)

PROCEDURE

Wear your safety goggles, apron, and gloves while performing this experiment.

Part I: Preparation of Oxygen Gas

1. Weigh out about 10 g of potassium chlorate, $KClO_3$, and 3 g of manganese dioxide, MnO_2. Pour both substances into a large pyrex test tube and mix gently until the mixture appears gray. Put a loose plug of glass wool about 2 cm into the tube. Clamp the test tube to a ring stand as shown in Figure 12–1. Stopper the tube with a one-hole stopper containing a right angle glass bend. The reaction mixture should not extend more than halfway up the tube.

    CAUTION: $KClO_3$ is reactive and should be handled carefully. It should never be mixed by grinding or be heated in contact with organic materials.

2. Fill a pneumatic trough with water. Connect a piece of rubber tubing from the trough drain to the sink to handle any overflow. Fill four wide-mouth bottles with water. Holding a glass plate over the mouth, invert each bottle and place it in the trough. Connect the test tube to the trough by a length of rubber tubing as shown in Figure 12–1. Put a piece of glass tubing bent in a right angle on the end of the rubber tubing that goes in the trough. When your set-up is complete, have it approved by your teacher.
3. (a) Gently heat the test tube containing the $KClO_3/MnO_2$ mixture. Heat the mixture evenly, being careful to avoid heating the stopper. (If the mixture smokes you are heating it too strongly.)
   (b) Allow the gas that comes off in the first few minutes to escape in order to clear the system of air. Then insert the tubing into a bottle.
   (c) Collect four bottles of oxygen gas. As the gas enters the bottle, the water inside will be displaced. Leave about 1 cm of water in each bottle. As each bottle is sufficiently filled, raise it just enough to slide a glass plate across its mouth. Holding the glass plate in place, remove the bottle from the trough and stand it upright. When the last bottle is full, take the tubing out of the water before removing the flame. Otherwise water will be drawn into the test tube as the gas inside the tube cools and contracts.

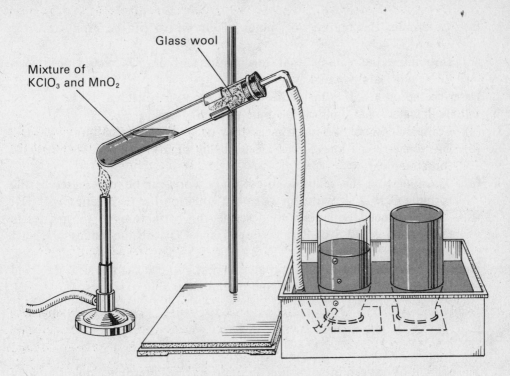

FIGURE 12–1 Apparatus for generation and collection of oxygen gas from $KClO_3$.

Part II: Properties of Oxygen Gas

1. Carry out a combustion reaction in pure oxygen by inserting an ignited wad of cotton or wooden splint into a bottle of oxygen. Observe the results and compare to combustion in air.

2. Obtain a small wad of steel wool. Hold the wool with tongs and heat it to glowing in a Bunsen burner flame. Quickly lower the glowing wool into a second bottle of oxygen and hold it there with the tongs. Observe the reaction. When the reaction is over, drop the residue into the water in the bottle to cool. Retrieve a piece of the residue and examine it. Compare its physical properties to those of the original steel wool. Test the solubility of the residue by attempting to dissolve it in water.

3. Place a piece of charcoal, about the size of a pea, on a deflagrating spoon. Heat the charcoal, which is essentially pure carbon, to glowing and lower it into the third bottle of oxygen. Wait a few minutes, then remove the spoon and any charcoal residue. Cover the bottle, and shake vigorously. Add 1 ml of a solution of $BaCl_2$ to the water. Record your observations.

4. Remove the glass plate from the last bottle of oxygen and sniff carefully to determine its odor. Oxygen is odorless. Record your conclusions about the sample you collected.

CALCULATIONS AND QUESTIONS

1. (a) Determine the number of moles of O_2 which can be produced from 10 g of $KClO_3$.
   (b) Determine the volume that this amount of dry O_2 would occupy at 25°C and 1 atmosphere.
2. Describe the odor and color of the oxygen gas you collected. Is the gas very soluble in water? Was your sample pure oxygen? Explain.
3. (a) Compare combustion reactions as they occur in air and in pure oxygen.
   (b) Oxygen is sometimes described as being highly inflammable. Criticize this statement.
4. (a) Assuming that the cellulose in cotton or wood can be represented by the formula $C_6H_{10}O_5$, write the balanced equation for its reaction with O_2.
   (b) Write balanced equations to describe the formation of the oxides in Procedures 2 and 3 of Part II. Fe_2O_3 and CO_2 are the oxides which are produced.
5. Based upon your observations, compare as best you can the physical state and solubility of a metallic oxide with that of a non-metallic oxide.

EXTENSIONS

1. The percent of oxygen in potassium chlorate may be determined quantitatively as an extension of this experiment. The procedure is as before except that the masses of $KClO_3$, MnO_2, and the test tube must be precisely determined.
   (a) In Procedure 1, Part I, weigh the empty test tube. Add the $KClO_3$ and weigh again. Then add the MnO_2 and weigh once more. All weighings should be made to the nearest 0.01 g. Continue with Procedure 1 as described.
   (b) After the oxygen collection is finished in Procedure 3, continue to heat the $KClO_3/MnO_2$ mixture with the tubing immersed. When the rate of oxygen evolution slows down, increase the intensity of the heat and heat strongly until no more gas is liberated. Keep the heat away from the stopper as much as possible during this step. Remove the hose from the water and then remove the flame from the mixture.
   (c) After the test tube is cool, remove the stopper and weigh the tube and contents. Calculate the mass of oxygen evolved in the reaction.
   (d) Assuming that all of the oxygen came from the potassium chlorate, calculate the mass percent oxygen in the compound.
   (e) Compare the experimental percent oxygen obtained above to the theoretical percent oxygen in $KClO_3$. Calculate your percent error.

PERIODICITY AND PREDICTIONS OF PROPERTIES

OBJECTIVES

1. To become familiar with the organization of the periodic table.
2. To study a periodic property.
3. To make predictions based on periodic properties.
4. To use the periodic table to assist in writing chemical formulas.

DISCUSSION

The periodic law states that many properties of the elements are periodic functions of their atomic numbers. A periodic function is one which goes through cycles, with maximum and minimum values at regular intervals. The atomic radius, melting point, and boiling point are periodic properties of the elements. The periodic table arranges the elements in order of increasing atomic number in such a way that the periodic nature of properties is made clear. This is done by placing in the same vertical column of the table those elements which appear at corresponding positions in the cycle of properties. For example, since the atomic radii of Li, Na, K, Rb, and Cs appear at maxima in the cycles for atomic radii as a function of atomic number, we would expect that these elements would be in the same column in the periodic table. These elements belong to a family, or Group, and have many other properties which would also indicate that they should be classified together. In general, properties of the elements in a Group tend to change gradually as one goes from the lightest to the heaviest atom in the Group. Within the framework of the periodic table properties vary according to a pattern as you move across the table in a cycle, or period, or up and down one of the Groups. If the pattern for the variation is known, it is often possible to predict a property of one element from the properties of elements which lie nearest that element in the table. Depending on the position of the element in the table, the best prediction will either be based on elements in the same row, or on elements in the same column, as the element for which the prediction is to be made.

In this experiment you will make some graphs of periodic properties against atomic number in order to discover the patterns the properties follow. You will then use those patterns to predict some properties of elements. In the last part of the experiment you will be given a blank periodic table and asked to assign elements to their proper positions in the table on the basis of their characteristic properties.

PRELIMINARY STUDY

1. Review Sections 7.3, 7.4, and 7.5 in the text.
2. Define ionization energy, atomic radius, and metallic character. How do these properties vary as one goes down a Group in the periodic table? Across a period?

PROCEDURE

1. Using the data in Table 13–1, plot the atomic radius of each element on the y-axis against its atomic number on the x-axis. Make the x-axis on the long side of the paper and choose the scales so that all the data will fit on the sheet. Connect each two consecutive points with a straight line. Label each peak with the symbol of the element.

TABLE 13–1   Atomic Radii of the Elements in nm (10^{-9} m)

ELEMENT	ATOMIC NUMBER	ATOMIC RADIUS	ELEMENT	ATOMIC NUMBER	ATOMIC RADIUS
H	1	0.037	K	19	0.231
He	2	0.050	Ca	20	0.197
Li	3	0.152	Sc	21	0.160
Be	4	0.111	Ti	22	0.146
B	5	0.088	V	23	0.131
C	6	0.077	Cr	24	0.125
N	7	0.070	Mn	25	0.129
O	8	0.066	Fe	26	---
F	9	0.064	Co	27	0.125
Ne	10	0.070	Ni	28	0.124
Na	11	0.186	Cu	29	0.128
Mg	12	---	Zn	30	0.133
Al	13	0.143	Ga	31	0.122
Si	14	0.117	Ge	32	0.122
P	15	0.110	As	33	0.121
S	16	0.104	Se	34	0.117
Cl	17	0.099	Br	35	0.114
Ar	18	0.094	Kr	36	0.109

2. Using the data in Table 13–2, plot the melting points of the Group 1 elements. Plot temperature on the y-axis and atomic number on the x-axis. The

atomic number scale should extend to francium (atomic number 87). Draw a
smooth curve through the data points. It will not be a straight line.

TABLE 13–2   Melting Points of the Group 1 Elements

ELEMENT	Li	Na	K	Rb	Cs	Fr
AT. NO.	3	11	19	37	55	87
MELTING POINT (°C)	186	98	64	39	---	---

3. Fill in the blanks in the periodic table provided you by your teacher with the
   letter symbol for each of the mystery elements whose properties are described.
   The idea is to identify the mystery element and to put its symbol in the
   proper place in the blank table. In doing this part of the experiment you may
   use the periodic table in your text and any information in the text. You may
   find the information in Chapter 7 to be particularly helpful.
   (a) A, B, C, and D belong to a family, the members of which are all gases.
       A is commonly used in advertising signs. B was first discovered on the
       sun and is used in weather balloons. C was used to make the first com-
       pounds of an element of this group. D is the family member which is
       present in the largest amount in air.
   (b) E and F are members of a family containing both gaseous and solid
       elements. E forms a diatomic molecule and is the major constituent of
       the atmosphere. Element F is a dangerous poison and a metalloid.
   (c) G, H, and I belong to a family of very active metals, all of which react
       with chlorine to produce salts with the general formula XCl. G is a
       member of the first period to contain 18 elements. H has the highest
       ionization energy of the family and I has the lowest.
   (d) J, K, and L belong to the same family and all are metals. A compound of
       J is a major component of bones and teeth. K is commonly used in flash
       bulbs for producing light. L is a radioactive element discovered by Marie
       Curie.
   (e) M is a gas and has some properties similar to the elements in both Group
       1 and Group 7. It is a unique element in this respect.
   (f) N, O, and P commonly form – 1 ions when they combine with metals.
       N is a liquid and O is a non-radioactive solid. P is the most chemically
       reactive of all the non-metals.
   (g) Q, R, S, and T are in different families but in the same period. Q is a
       gas used for water purification. R is a yellow non-metallic solid. S is a
       metalloid used in transistors. T is a metal of low density used in aircraft
       construction.
   (h) U, V, W, and X are all transition elements. U is an excellent conductor
       of heat and electricity and is commonly used in wiring and cookware.
       V is the only metal that is a liquid at room temperature. W is the metal
       which is produced in the largest quantity. Although once used in coins,
       X is now used mostly in expensive jewelry.
   (i) Y is an actinide fuel used in nuclear reactors. Z is the actinide named
       for the creator of the Periodic Table.

CALCULATIONS AND QUESTIONS

1. (a) Describe any regularities that are present in the graph of atomic radius *versus* atomic number. Which elements occupy the peaks in the cycles? Are the periods, or cycles, of the same length?

   (b) In Table 13–1 we left the atomic radii of Mg and Fe blank intentionally. The values are actually known, but we would like to have you predict them on the basis of the pattern of properties on the graph. Note that the part of the cycle from Li (3) to B (5) is repeated in the part of the cycle from Na (11) to Al (13). We would expect, then, that the radius of Mg would have a value that would maintain the pattern observed between Li and B. On that basis, predict the radius of the Mg atom.

   (c) As part of the last cycle on the graph, which starts at K (19), we have the first series of transition metals. By assuming that the Fe (26) atom would maintain the pattern set by atoms near it in atomic number, predict the radius of the Fe atom.

2. (a) In 1 we predicted the radii of Mg and Fe on the basis of radii of elements which lie to the left and right of those two elements in the periodic table. This method for prediction is most successful when the elements in question lie within a cycle and not at an end. If we attempt to predict the radius of an element at the end of a cycle, such as Na, from the radii of elements to the left and right of Na in the periodic table, we are doomed to failure. In such cases it is useful to plot the properties of the elements in a Group against atomic number. In the graph made in Procedure 2 we did this when we plotted the melting points of the Group 1 elements. Comment on the curve obtained. Is it of a form you would expect for elements within a Group?

   (b) The graph obtained in Procedure 2 can be used to predict the melting points of Cs (55) and Fr (87), which are the other two elements in Group 1. Here we need to extend, or extrapolate, the curve out to atomic number 87, maintaining the trend established by the lighter elements in the Group. By making this extrapolation, predict the melting points of Cs and Fr. Again, compare your values with those given in the literature. Perhaps the best source of such information is the Handbook of Chemistry and Physics.

3. Using the periodic table prepared in Procedure 3, predict the formulas for compounds of the following elements: (Use the mystery letters in the formulas.)

   (a) M and P   (c) E and M   (e) A and G
   (b) T and Q   (d) I and R   (f) J and N

   In predicting the formulas you may find the following formulas of known substances to be useful:

$$AlBr_3 \qquad HCl \qquad PH_3$$

$$Li_2O \qquad BaCl_2 \qquad NaI$$

The principle to be used here is discussed in Section 7.3 in the text.

EXTENSIONS

1. Using the data from Table 7–1 of Chapter 7 of your text, plot either melting point or boiling point for the first 20 elements. Describe the regularities in the graph. Is the property you plotted a periodic property?

SOME CHEMICAL PROPERTIES OF THE MAIN GROUP ELEMENTS

OBJECTIVES

1. To become familiar with the properties of some elements in Groups 1, 2, 6, and 7.
2. To observe trends in the properties of the elements in a Group.
3. To contrast the properties of metals with those of non-metals.

DISCUSSION

The main group elements belong to Groups 1 to 8 in the periodic table. They do not include the transition metals, the lanthanides, or the actinides. Among the main group elements we find the most reactive metals, all the non-metals, and the metalloids.

The Group 1 elements are all soft, silvery, highly active metals. They react quickly with air, acquiring a rather hard crusty oxide coating. In water these metals all react spontaneously, evolving hydrogen and forming the metallic ion and hydroxide ion in solution. The reaction of sodium is typical:

$$2\ Na(s) + 2\ H_2O(1) \rightarrow 2\ Na^+(aq) + 2\ OH^-(aq) + H_2(g)$$

Group 2 elements are harder, somewhat less reactive metals than those in Group 1. They oxidize to some extent in air, but are much less violent in reaction with water than are the metals in Group 1. The reaction products with water are similar to those with Group 1, except that the ion produced has a charge of +2. Group 2 hydroxides are much less soluble than those of Group 1, so when the metals in Group 2 react with water one may observe formation of the solid hydroxide.

The properties of the Group 6 elements vary greatly within the group. Oxygen is a gas, sulfur a nonmetallic solid, selenium and tellurium are metalloids, and polonium is a radioactive metal. By comparison with the metals in Groups 1 and

2, sulfur is nonreactive toward air and water. Sulfur will burn if heated in air, forming sulfur dioxide. At high temperature it will react with metals, forming sulfides. The element normally exists as a crystalline solid, containing S_8 ring molecules. The solid melts at about 113°C to a light colored liquid. If this liquid is heated, the S_8 rings open up and the chains link to form a polymer. As a result of this change, the liquid becomes dark red-brown in color, and its viscosity increases greatly. If the hot viscous liquid is cooled quickly in water it forms a rubbery material, called plastic sulfur, very unlike the original yellow solid. If the polymeric sulfur is allowed to stand, it reverts in a few hours to ordinary sulfur.

The elements in Group 7 all normally exist as diatomic molecules. Fluorine and chlorine are gases, bromine is a liquid, iodine and astatine are solids. The elements are more soluble in water than is sulfur, but do not show the strong reaction with water exhibited by the active metals. The elements dissolve more readily in organic liquids than in water and often show characteristic colors in such liquids. In chemical reactions all of the elements in Group 7 tend to remove electrons from other species, becoming negative ions with a −1 charge. Chlorine is more powerful in this regard than bromine or iodine, and will remove electrons from bromide or iodide ions. The reaction usually occurs in water solution; with bromide ion the reaction is:

$$Cl_2(aq) + 2\ Br^-(aq) \rightarrow 2\ Cl^-(aq) + Br_2(aq)$$

Bromine and iodine are prepared commercially by the above reaction. In the laboratory Cl_2, Br_2, and I_2 can be made by the reaction of the halide (Group 7) ion with MnO_2 in the presence of high concentrations of H^+ ion. With bromide ion the reaction is:

$$2\ Br^-(aq) + 4\ H^+(aq) + MnO_2(s) \rightarrow Br_2(aq) + Mn^{2+}(aq) + 2H_2O$$

In this experiment we will carry out many of the reactions we have discussed here. These reactions illustrate the wide variety of properties of the elements and should increase your familiarity with the chemical behavior of some common substances.

PRELIMINARY STUDY

1. Review Chapter 8 in the text.

PROCEDURE

Part I.   Reactions of elements in Group 1 with air and water.

Since the metals in Group 1 are extremely reactive, your instructor will demonstrate the behavior of these elements. These metals are potentially very dangerous, so no one who is not thoroughly familiar with their properties should attempt to work with them. Your teacher will perform the next three steps.

1. Remove pieces of lithium, sodium, and potassium from their protective liquids. Slice each metal with a knife to expose a fresh surface. Note the behavior of the surface and compare the fresh surface to the surface of the large piece of that metal.

2. Add a small piece of each alkali metal to separate 400 ml beakers half filled with water. Note the reaction which occurs and compare the reactivities of the three metals.

3. To each solution add a few drops of phenolphthalein. This substance, which is called an acid-base indicator, turns red in solutions containing appreciable amounts of OH^- ion.

Wear your safety goggles, apron, and gloves during the rest of this experiment. Perform Parts II, III, and IV under a fume hood.

Part II. Reaction of a Group 2 element with water.

1. To a few milliliters of water in a test tube add a small piece of calcium metal. Observe the reaction and compare it to those of the Group 1 metals, with respect to both the intensity of the reaction and the nature of the products.

2. Add a few drops of phenolphthalein to the solution. Note your observations.

Part III. Some properties of sulfur.

1. To a test tube containing a few ml of water add a small piece of sulfur. Contrast the behavior of sulfur in water with that of the Group 1 elements.

2. Obtain a wide-mouth bottle and to it add about 10 ml of water. Put about a gram of sulfur on a deflagrating spoon. Ignite the sulfur with the flame from a Bunsen burner and quickly lower the spoon into the bottle. When the sulfur has finished burning, remove the spoon, stopper the bottle, and shake it. Then add 1 ml of a solution of $Pb(NO_3)_2$ to the liquid in the bottle. Record all your observations during this procedure.

3. Put a small piece (about 2 g) of sulfur in a large test tube. Using a test tube holder, slowly heat the tube in a Bunsen flame. Note any changes in color and viscosity that occur on melting and on further heating. Continue heating until the sulfur is near the boiling point (about 450°C). (Do not ignite the sulfur.) Clamp the tube vertically and lower a clean penny on a piece of cellophane tape into the vapors. After a few moments, remove the coin and compare the untaped and taped surfaces. Then put a glass stirring rod into the molten liquid, remove some of the hot sulfur and quench it in a beaker of water. When it has cooled, remove the product from the rod and compare its properties with those of the sulfur you started with. Observe and record the properties of the sulfur in the large tube as it cools and solidifies.

Part IV. Some properties of the halogens.

1. *Preparation of iodine.* Place about 2 grams of sodium iodide, NaI, and 1 gram of manganese dioxide, MnO_2, in a 250 ml beaker. Add about 10 ml of dilute sulfuric acid, 3 M H_2SO_4, and stir the mixture. Put an evaporating dish containing cold water (or, better, an ice cube) on top of the beaker. Set the beaker and dish on a wire gauze and iron ring and heat very gently with a Bunsen burner. Iodine vapor will be produced in the reaction and will condense on the cold surface of the dish. After a minute or two, when an appreciable amount of iodine has crystallized on the dish, stop heating and let the covered beaker cool for a few minutes. Then remove the evaporating dish and scrape the crystals on to a piece of filter paper with a spatula.

CAUTION: Do not overheat. These procedures should be done under a fume hood. Wear your safety goggles, apron, and gloves at all times.

2. *Properties of iodine.*
   (a) Add a few of the iodine crystals to two separate test tubes, one containing about 2 ml of water and the other about 2 ml of trichloroethane, CCl_3CH_3, an organic solvent. Compare the solubility of iodine in the two liquids. Note the color of iodine in the solutions.
   (b) Put a few iodine crystals in a small test tube. Stopper the tube and warm the bottom of the tube with your hand. Record your observations.
3. *Reactions of chlorine with the halide ions.* To a small test tube add about 2 ml of a dilute sodium chloride solution (0.1M NaCl). To another tube add 2 ml of a sodium bromide solution (0.1M NaBr), and to a third tube add 2 ml of a sodium iodide solution (0.1M NaI). Then to each tube add 5 drops of a solution of Cl_2 in water (chlorine water). Stopper each tube and shake it thoroughly. Note any reactions. To each tube then add about 1 ml CCl_3CH_3, stopper, and shake. CCl_3CH_3 does not mix with water and is much denser than water. Record all your observations.

QUESTIONS

1. (a) Based on their reactions with water, list Na, K, and Ca in order of their decreasing chemical reactivity. Using the trends observed, predict how Mg would compare in reactivity to Na, K, and Ca.
   (b) Write the equation for the reaction of potassium with water.
   (c) What evidence do you have that OH^- ions are produced in the reactions of elements in Groups 1 and 2 with water?
   (d) What can you say about the solubility of $Ca(OH)_2$ in water?
2. Compare the behavior of sulfur and iodine in water to that of the Group 1 metals. Which elements undergo chemical changes? Which undergo physical changes? Which don't undergo anything?
3. (a) Write the equation for the reaction of sulfur with air.
   (b) Describe the product of the reaction described by the equation just written, as to odor, physical state, and solubility in water.

(c) Explain why the sulfur that was quenched in water did not have the properties of ordinary sulfur.

4. (a) Write the equation for the reaction by which iodine is made from sodium iodide, manganese dioxide, and sulfuric acid.

(b) In the reaction, what is the role of the NaI? the sulfuric acid?

(c) What can you say about the vapor pressure of iodine compared to that of other solid elements?

(d) Is iodine more soluble in water or CCl_3CH_3? Can you suggest why there might be a difference in solubility in the two solvents?

(e) Write an equation for the reaction of chlorine water with a solution of sodium iodide.

(f) A solution contains one salt. That salt may be NaCl, NaBr, or NaI. How could you find out which salt was present?

EXTENSIONS

1. Generation of chlorine gas.

(a) Commercial bleaching solutions are usually 5% sodium hypochlorite, NaClO. Place 20 ml of a standard bleaching solution in a wide-mouth bottle. Add about 10 ml of dilute hydrochloric acid, 6M HCl, and stopper the bottle immediately. Note the color of the chlorine gas which is produced.

CAUTION: Chlorine is a poisonous gas. This reaction should only be carried out in a hood.

(b) Moisten a piece of colored paper or cotton cloth. Momentarily remove the stopper and hang the colored material in the gas in the bottle, holding it there by replacing the stopper. After several minutes check the color of the material.

UNDERSTANDING ELECTRON CHARGE DENSITY DIAGRAMS

OBJECTIVES

1. To determine the distribution of dart hits about a bullseye.
2. To obtain and interpret probability information.
3. To compare the results of the dart experiment to the quantum theory prediction of the electron density around an atomic nucleus.

DISCUSSION

When scientists study the motion of small particles like electrons, they find that these particles do not seem to behave like more familiar larger particles such as billiard balls or bicycle wheels. With the larger particles one can determine the path the particle will follow, given its initial speed and direction. This does not seem to be possible with electrons. In order to deal with the rather different properties of electrons, a science called quantum mechanics was developed, to be used instead of classical mechanics, which works well for larger particles.

According to quantum mechanics the best we can do in our experiments is to find the likelihood that an electron will be at a given point in an atom. Say we did a hundred experiments to locate an electron in an atom. We would find the electron in one hundred places, some of which were closer to the nucleus than others. Perhaps in ten of the experiments the electron would be found in a certain region in the atom. We would say, then, that the probability of finding the electron in that region is 10/100, or 0.10. This kind of information, while useful, is far less specific than knowing the path of the electron. Quantum mechanics tells us we can find probabilities of the sort just mentioned, but not the path an electron will follow.

In order to help make clear how one obtains and interprets data on electron positions, we are going to do an experiment which will furnish very similar data. In our experiment we will drop darts on to a target like the one shown in Figure 15–1. The target consists of ten concentric rings one cm wide. If we drop darts on to such a target, we find that not all of the darts hit the bullseye. There will be a scatter of hits over the target, with more hits near the bullseye than far from it (if you are a good dart dropper, that is).

From the pattern of hits we can determine the probability that a dart will hit in a given ring; we will call this the hit probability. The hit probability essentially tells us the likelihood that a dart will hit at any given distance from the bullseye. We can also find a related quantity, which we call the hit density, which is the probability that a dart will hit a particular unit area located on a given ring on the target. If you think about this a moment, you can see that the hit density at a given distance from the bullseye will have to equal the hit probability at that distance divided by the area of the ring at that distance. If, for example, 16 out of 100 darts hit in ring number 3 in Figure 15–1, we can say that the probability that the next dart will hit that ring is 16/100, or 0.16. That is the hit probability for that ring. Since the area of that ring is 16 cm², the hit density would equal 0.16/16, or 0.01. This means that if we placed a square 1 cm on an edge anywhere on ring number 3, the odds are that 1 dart out of the next 100 would hit somewhere on that square. You could also say that the hit density at any given radial distance is the probability that a dart will strike a particular unit area at that distance from the bullseye. If you are interested in problems of this sort, you might try to guess what the experiment will tell us about how the hit probability and the hit density vary as we move from the bullseye out to the more distant rings.

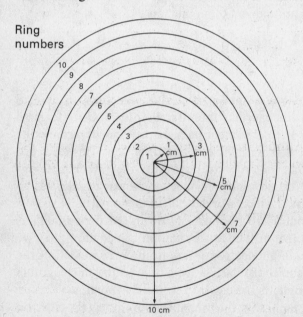

Ring numbers

FIGURE 15–1 A two-dimensional target for determining the probability of dart hits around a bullseye.

The results of quantum mechanics calculations of electron positions are very much like those from the dart experiment. One result, shown in Figure 15–2A is an electron probability graph. This graph is for the 1s electron in a hydrogen atom. It tells us the probability that the electron will be at any given distance from the nucleus (that is, will be found in a thin spherical shell at that distance). The electron probability is like the hit probability, except that since atoms are spherical rather than flat, the probability must be applied to a spherical shell instead of a ring. In Figure 15–2A we see that in the H atom the electron is most likely to be found at a distance of about 0.05 nm from the nucleus. It is not likely to be far from the nucleus, nor is it likely to be at the nucleus.

The other result from quantum mechanics is shown in Figure 15–2B, where we have plotted the electron charge density in the H atom as a function of distance from the nucleus. This graph gives the probability that an electron will be found in a given small volume located at a particular distance from the nucleus. The electron charge density curve is like the hit density curve, and essentially tells us the likelihood of an electron being found at any given point in the atom. From the graph we see that the charge density is largest at the nucleus. If we had to select a small region in the atom where we would be most likely to find the electron, the best place to put that region would be at the nucleus. If we put that region, of fixed volume, anywhere else, we would be less likely to find the electron there.

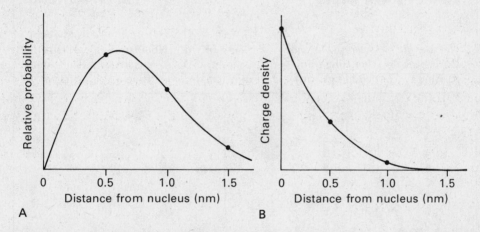

FIGURE 15–2   Graphs showing electron probability and electron charge density for a 1s electron in a H atom.

Frequently electron charge densities in atoms are presented in diagrams like that in Figure 9.5 in the text, where the electron charge cloud in a H atom is represented. The heavier density at the center of the atom reflects the high charge density at that point. Electron orbitals used to indicate electron positions are really charge clouds showing electron density in the atom.

PRELIMINARY STUDY

1. Review the quantum theory model of the atom discussed in Section 9.2 of the text.

PROCEDURE (To be done in pairs)

1. Obtain a target, masking tape, plywood board, and a dart. The concentric circles on the target have the radii shown in Figure 15–1. Tape the target to

the center of the board and place it on the floor. Stand on a stool and extend your arm so that it is about two meters directly above the target. Drop the dart (do not throw) to the target below in such a way as to try to hit the bullseye. The second student should retrieve the dart and mark the position of the hit with a small x. Repeat this procedure 99 times for a total of 100 drops. (Students should reverse positions at the half-way point.) Don't count drops which fall outside the largest circle. Detach the target and return the dart and board.

2. Count the number of hits in each concentric ring and record in the data table. Divide the number of hits for each ring by its area to determine the hits per cm^2.

SAMPLE DATA TABLE   **EXPERIMENT 15**

Number of Concentric Ring	Average Distance of Ring from Bullseye (cm)	Area of Concentric Ring (cm^2)	Number of Hits in Ring	Number of Hits per Unit Area ($hits/cm^2$)
1	0.5	3.1		
2	1.5	9.4		
3	2.5	16		
4	3.5	22		
5	4.5	28		
6	5.5	35		
7	6.5	41		
8	7.5	47		
9	8.5	53		
10	9.5	60		

CALCULATIONS AND QUESTIONS

1. (a) For each ring, plot the number of hits in the ring on the y-axis against the average distance of the ring from the center on the x-axis. Beginning at the 0,0 point, draw a smooth curve through the plotted points.

   (b) This curve represents the hit probability as a function of the average radius of the ring. What is the probability a dart will hit in ring 4? From your graph, at what distance from the bullseye will a dart be most likely to hit?

   (c) Compare this curve to the electron probability curve in Figure 15–2A. State in words what the graph in Fig. 15–2A tells us about the electron.

2. (a) For each ring, plot the hit density ($hits/cm^2$) on the y-axis against the average distance of the ring from the bullseye on the x-axis. Draw a smooth curve through the plotted points, and extend the curve to x equals 0.

   (b) State in words how the probability of a hit in any given unit area on the target varies with the distance of that area from the bullseye. Where would you put a square, 1 cm on edge, on the target to maximize the likelihood of its being hit?

(c) Compare the hit density curve to the electron charge density curve in Figure 15-2B. Where in the atom would you put a region of fixed small volume if you wished to maximize the chance of finding the electron inside that volume?

3. (a) Would you expect the radius of maximum probability for a dart hit to be the same for all students? (Graph 1) Why?

(b) Would you expect the most probable radius of the 2s electron in a lithium atom to be the same as that for the 1s electron in a hydrogen atom? Why?

EXTENSIONS

1. (a) In making the graph in Procedure 1, we did not consider the effect of the direction your arm was pointing when the dart was dropped. By examining the target again, check to see if there was a directional effect. (You can do this by drawing a line through the bullseye parallel to your arm direction, and another line through the center perpendicular to the first line. Count hits in each quadrant for each concentric ring.) Comment on the directional effect.

(b) How does the directional effect in the target game compare to the directional effect in the electron probability graph in Figure 15-2A?

ANALYSIS OF A HYDRATE

OBJECTIVES

1. To find the formula of a hydrate.
2. To become more familiar with the properties of hydrates.
3. To review mass relationships in chemical reactions.

DISCUSSION

When an ionic solid is crystallized from water solution, the crystal which forms often contains chemically-bound water molecules. The number of moles of water per mole of ionic substance is usually an integer. Compounds of this sort are called hydrates. Among the commonly encountered hydrates are the following:

$$CuSO_4 \cdot 5H_2O \qquad Na_2HPO_4 \cdot 12H_2O$$

$$MgSO_4 \cdot 7H_2O \qquad BaCl_2 \cdot 2H_2O$$

The water in a hydrate is bound loosely, and so is relatively easily removed by heating. Most hydrates lose their water of hydration at temperatures slightly above 100°C. Sometimes the water is liberated in stages, with one or more lower hydrates being observed during the heating process. Thus, $CuSO_4$ may also be prepared with 3 moles of H_2O or 1 mole of H_2O per mole of ionic solid. If all the hydrated water is removed, as it will be if the solid is heated sufficiently, the ionic solid is said to be anhydrous (without water).

Given the mass of a sample of the hydrate and the mass of anhydrous salt of known formula obtained on heating, it is easy to find the formula of the hydrate. One simply needs to determine the number of moles of water per mole of anhydrous compound in the hydrate:

mass of water in hydrate = mass of hydrate − mass of anhydrous solid

$$\text{no. of moles of water in hydrate} = \frac{\text{mass of water in hydrate}}{\text{mass of one mole of water}}$$

$$\text{no. of moles of ionic compound in hydrate} = \frac{\text{mass of anhydrous solid}}{\text{mass of one mole of anhydrous solid}}$$

$$\frac{\text{no. of moles of water in hydrate}}{\text{no. of moles of anhydrous solid}} = \text{mole ratio } H_2O\text{:anhydrous salt} = X$$

$$\text{formula of hydrate} = \text{formula of anhydrous salt} \cdot XH_2O$$

PRELIMINARY STUDY

1. Review Section 10.5 in the text.
2. Practice Problem: In an experiment 3.00 g of the hydrate $BaCl_2 \cdot XH_2O$ is heated to remove the water. The remaining anhydrous salt has a mass of 2.56 g. Determine the mole ratio of water to salt and the formula of the hydrate. (Ans: $BaCl_2 \cdot 2H_2O$)

PROCEDURE

Wear your safety goggles, apron, and gloves while performing this experiment.

1. Obtain a clean crucible and cover. Place the crucible and cover on a clay triangle, with the cover slightly ajar as shown in Figure 16–1. Heat with a bunsen burner for 3–4 minutes, gently at first and then strongly. This will drive off

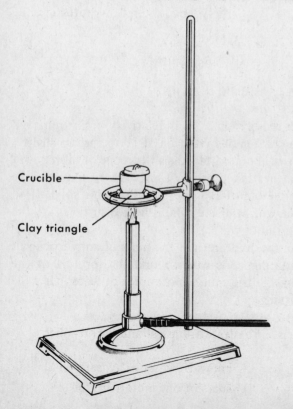

Crucible

Clay triangle

FIGURE 16–1 Apparatus used for removing the water from a hydrate.

any water that is adsorbed on the walls of the crucible.

2. Allow the crucible and cover to cool on the triangle or heat resistant pad.
3. When the crucible and cover are cool, weigh them on a balance. Make this and all successive weighings to ± 0.01 g.
4. Put about 3 g of the hydrate sample in the crucible, replace the cover and weigh.
5. Place the crucible on the clay triangle on a ring stand. With the cover on the crucible slightly ajar as before, heat the crucible and its contents, gently at first and then strongly, for approximately 8–10 minutes. The hydrate will probably have a different appearance when the water has been driven off.
6. Slide the lid to completely cover the crucible and allow it to cool. When they are at approximately room temperature, weigh the crucible, cover, and contents and record the mass.

SAMPLE DATA TABLE   **EXPERIMENT 16**

Anhydrous salt formula   _____

Mass of empty, dry crucible and cover   _____ g

Mass of crucible, cover and hydrate sample   _____ g

Mass of crucible, cover, and anhydrous sample   _____ g

CALCULATIONS AND QUESTIONS

1. Using the experimental data obtained, determine:
   (a) the mass of water lost by the hydrate.
   (b) the number of moles of water lost.
   (c) the mass of the anhydrous salt present.
   (d) the number of moles of anhydrous salt present.
2. (a) Using the answers obtained in 1(b) and 1(d), obtain the ratio:

$$\frac{\text{moles of water}}{\text{moles of anhydrous salt}}$$

   (b) Write the formula of the original hydrate by rounding off the above ratio to the nearest whole number or simple fraction.
3. Consider what effect the following errors would have on the ratio of moles of water to moles of salt in Question 2(a):
   (a) the student did not drive off all of the water from the hydrate.
   (b) the student used a damp crucible and did not dry it before adding the hydrate.
   (c) after the last heating the student allowed the crucible and contents to cool overnight in the air before weighing.

EXTENSIONS

1. Some anhydrous ionic substances tend to absorb water from the air to form hydrates. Such materials are said to be hygroscopic. In some cases the attraction of the anhydrous solid for water is so strong that the solid can be used as a desiccant. A desiccant is used in a desiccator to dry the enclosed air and hence keep the contents of the desiccator from absorbing water vapor. A common desiccant is anhydrous calcium chloride.

   (a) Put about 3 g of anhydrous calcium chloride into a previously weighed 50 ml beaker. Weigh the beaker and contents. Let the sample remain in the air until the next laboratory period, and then reweigh the beaker and its contents.

   (b) Calculate the number of moles of water absorbed per mole of calcium chloride. Would you say the sample is now a hydrate?

COVALENT BONDING AND MOLECULAR STRUCTURE

OBJECTIVES

1. To develop skill in making and interpreting molecular models.
2. To draw Lewis structures for molecules and ions.
3. To use Lewis structures and molecular models to predict molecular geometry and polarity.

DISCUSSION

One of the important ways in which atoms can become chemically bonded is by covalent bonds. A covalent bond contains electrons which are shared by the bonded atoms. By sharing electrons atoms can acquire more stable structures. A particularly stable structure is the one in which all of the atoms have a share in eight electrons (except for hydrogen atoms, which have a share in two electrons). Molecules which have such structures, and there are a great many that do, satisfy the octet rule. All of the molecules we will study in this experiment obey the octet rule. Their electronic structures are most easily obtained by drawing their Lewis structures, using the approach described in Section 11.2 in the text.

In molecules which obey the octet rule the eight electrons on each atom are arranged in four pairs. Each pair of electrons may be present either in a covalent bond or as a non-bonding pair that is not attached to another atom. If all of the four electron pairs around an atom are either in single bonds or in non-bonding pairs, the four pairs of electrons will be oriented toward the corners of a tetrahedron, at the center of which is the atom being studied. In this arrangement the electron pairs are as far away from each other as possible, and so satisfy the electron pair repulsion rule.

The geometry of a molecule depends on the arrangement of its atoms, and not its electrons. If a molecule contains four atoms bonded to a central atom, as in CH_4, or $CHCl_3$, the four atoms will be bonded to the central atom by single bonds consisting of electron pairs, and since the electron pairs are arranged tetrahedrally, the molecule itself will be tetrahedral. If the central atom is bonded to three atoms, as in NH_3, three of the four pairs around the central atom are used in bonds, and the fourth pair is non-bonding. The NH_3 molecule is not tetrahedral, since the atom that would complete the tetrahedron is missing; we describe this molecule as being pyramidal. In a similar way, we conclude that the H_2O molecule is bent, and the HF molecule is linear.

To determine the geometry of molecules containing a double or triple bond, one can make a molecular model by the approach described in this experiment. A model constructed so as to have a satisfactory Lewis structure will have the geometry of the molecule it represents. There are other ways to obtain molecular geometry, but the model method is both simple and reliable.

From the geometry of a molecular model one can determine whether a molecule is polar or nonpolar. A bond between unlike atoms will always be polar, which means it will have a + end and a − end. A diatomic molecule containing such a bond will also be polar. The HF, CO, and HCl molecules are all polar. N_2 and O_2 are nonpolar, because both ends of those molecules are equivalent. A polyatomic molecule may be nonpolar even though it contains polar bonds. In such a molecule, the charge distributions in the bonds interact in such a way that the molecule has no + and − ends. The CH_4 molecule is nonpolar because of this effect. The ABA molecule will be nonpolar if it is linear, and polar if it is bent.

In this experiment we will make ball and stick models for some very common molecules. The geometry of the electron pairs will be properly established by the model, so the geometry of the molecules will be that of the model. From the molecular geometry obtained from each model you should be able to determine the polarity or nonpolarity of the molecule it describes.

PRELIMINARY STUDY

1. Review Sections 11.2, 11.3 and 11.4 in the text.
2. Get a head start on the experiment by doing Procedure 1 before the lab period.

PROCEDURE

1. (a) List the following molecules in the first column of your data table. (Leave about a 3 cm space between each formula.)

CH_4	H_3O^+	N_2	C_2H_2
CH_2Cl_2	HF	Cl_2	SO_2
CH_4O	NH_3	C_2H_4	SO_4^{2-}
H_2O	H_2O_2	CH_2O	CO_2

    (b) Draw the Lewis structure of each molecule.
    (c) Predict the geometry of each molecule as best you can.
2. (a) Obtain a ball and stick molecular model kit. These kits contain balls for atoms, sticks for single bonds, and springs for multiple bonds. The

molecules for which we will make models contain atoms which obey the octet rule. Such atoms, in general, have four electron pairs and are represented by the balls which have four drilled holes. Hydrogen atoms in molecules which obey the octet rule share one electron pair and are represented by balls having a single hole.

(b) Construct the molecular model for each of the molecules in Procedure 1(a). Use sticks for single bonds and non-bonding electron pairs. Use springs for the multiple bonds. Use the same general method in assembling your model as you did in developing its Lewis structure. The completed model should satisfy the octet rule, so each hole should contain a stick or a spring. Compare the geometry of the model with the geometry you predicted in Procedure 1(c). Describe the molecular geometry in the data table.

(c) Examine the symmetry of each molecular model. Predict whether the molecule having that symmetry would be polar or nonpolar.

3. After completion of Procedure 2 have your data table checked by your instructor. If your results are satisfactory, you will be assigned a set of "unknowns." Record their formulas in the data table and build each molecular model. Complete the data table for each unknown, stating its Lewis structure, geometry and polarity.

SAMPLE DATA TABLE			EXPERIMENT 17
Molecule	Lewis Structure	Geometry	Polarity

EXTENSIONS

1. For some molecules with a given molecular formula, it is possible to satisfy the octet rule with two, three, or more different arrangements of the atoms. For example, the Lewis structure of the $C_2H_4Cl_2$ molecule can be drawn in two different ways:

H H   H H
| | | |
:Cl – C – C – H   :Cl – C – C – Cl:
| | | |
:Cl: H   H H

In one molecule the two Cl atoms are attached to the same carbon atom, while in the other, each carbon atom is bonded to a chlorine atom. (Note that there is free rotation around the C – C bond. This makes the three other bonding positions on each C atom equivalent. This will be very apparent if you make a model of the molecule.) Molecules with the same formula but different atomic arrangements are called isomers. $C_2H_4Cl_2$ exists in two isomeric forms.

Find all the isomers of $C_2H_2Cl_2$. Draw the Lewis structure, build the model, and describe the geometry and polarity of each isomer. (Around carbon-carbon double bonds there is no rotation. There are three isomers with the formula $C_2H_2Cl_2$.)

SOLID-LIQUID PHASE CHANGES

OBJECTIVES

1. To determine the melting point and the heat of fusion of an unknown solid.
2. To plot a time-temperature graph of a phase change and interpret the energy changes that occur.

DISCUSSION

Within a crystalline solid the particles, be they atoms, molecules, or ions, are arranged in a regular repeating lattice. At any temperature the particles undergo very rapid, small vibrations about their positions in the lattice. If the solid is heated, its temperature goes up, and the vibrations increase in intensity. Finally, at some temperature, which is called the melting point, the vibrations become large enough to overcome the forces that maintain the lattice structure, and the solid melts to a liquid. If the solid is pure, the temperature remains constant at the melting point during the entire melting process, from the time when melting begins to the time when the sample has all been converted to liquid. Once the solid is all melted, further heating will raise the temperature of the liquid.

If a liquid is cooled, the process we have described is reversed. The temperature of the liquid drops until we get to the freezing point; at that point solid appears, the temperature holds constant until the liquid is completely converted to solid; then the temperature drops again. The temperature at the melting point is the same as that at the freezing point, and is usually referred to as the melting point no matter which change is occurring. In this experiment you will determine the melting point of a pure substance by melting it and then cooling the liquid until it is completely converted to solid.

When a solid melts, energy must flow into it to break down the crystalline lattice. This energy is usually in the form of heat and is called the heat of fusion, ΔH_{fus}:

$$\text{Solid at M.P.} \rightarrow \text{Liquid at M.P.} \qquad \text{Heat effect} = \Delta H_{fus}$$

This means that the energy of a liquid is always higher than that of the solid from which it can be made. When a liquid freezes, heat is released to the surroundings. The change is completely reversible. It takes the same amount of heat to melt a solid as is released by the liquid when it solidifies. In the second part of this experiment you will measure the heat of fusion of a pure solid by finding the heat that is released by the liquid when it solidifies.

The melting point of a pure substance is one of its characteristic properties. Another such property is its heat of fusion per mole or per gram. In Table 18–1 we have listed the melting points and heats of fusion of several substances. These two properties can be used to identify a solid, and in this experiment we will use them both to establish the identity of an unknown substance with which you will be working.

TABLE 18–1   Melting Points and Heats of Fusion
of Selected Substances

SUBSTANCE	MELTING POINT (°C)	HEAT OF FUSION (CAL/G)
phenol	41	28.7
lauric acid	44	43.7
urethane	49	40.9
thymol	51	27.5
p-dichlorobenzene	53	29.1
myristic acid	54	47.5
nitronaphthalene	57	25.4
stearic acid	69	47.5
naphthalene	80	35.1

PRELIMINARY STUDY

1. Review Section 12.1 in the text.
2. Practice Problem: Gallium metal melts at 29°C. A tube containing 8.0 g of liquid gallium was immersed in a water calorimeter when the first crystal was seen to form. The temperature of the calorimeter, which contained 100 g of water, increased from 24.3°C to 25.8°C while solidification took place. Determine the heat of fusion of gallium in calories per gram. (Ans: 19 cal/g)

PROCEDURE (To be done in pairs)

Part I: Determination of the melting point of a solid

1. Obtain a 400 ml beaker, two styrofoam coffee cups, a thermometer, and a large test tube containing a sample of unknown solid. Fill the beaker about three-fourths full with tap water. Heat the water in the beaker to near boiling.
2. While the water is heating, pour the unknown solid into a clean beaker and

weigh the empty test tube. Add about 15 g of the unknown to the tube and weigh again. Place the test tube containing the weighed unknown into the water bath. As the solid begins to melt, put the thermometer into the tube and stir carefully. Note the approximate temperature at which melting occurs. Continue warming the liquid until the temperature is about 10° higher than the approximate melting point. Then turn off the Bunsen burner and leave the thermometer and the tube in the bath. While the solid is being melted, assemble a calorimeter by nesting the two styrofoam cups and supporting them in a 250 ml beaker. Add 150 ml of room temperature water to the inner cup.

3. At this point one student should elect to read temperature and the other to measure time and record data. Remove the test tube and melted unknown from the bath and clamp it to a ring stand as in Figure 18–1. Stir the liquid gently with the thermometer as it cools. When the first crystals appear, begin timing, record the temperature, and put the test tube in the water in the calorimeter. The level of the unknown should be well below the water level. Stir the sample gently as it solidifies, without touching the wall of the tube. The timekeeper signals for a temperature reading every 30 seconds and records it in the data table. The temperature should remain nearly constant as the solid forms. When the sample is mostly solid, stop stirring. Note the time when the temperature of the solid begins to fall relatively rapidly; at that time all the liquid has solidified. Continue to take readings until the temperature is about 10° below the freezing point. Start to reheat the water in the bath.

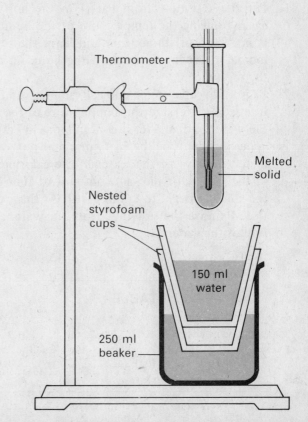

FIGURE 18-1 Apparatus used to observe the cooling behavior of a liquid.

Thermometer

Melted solid

Nested styrofoam cups

150 ml water

250 ml beaker

SAMPLE DATA TABLE   **EXPERIMENT 18 (Part I)**

mass of test tube _____ g   mass of test tube and unknown _____ g

mass of unknown _____ g   unknown number   _____

TIME (MIN)	TEMP (°C)	
0	_____	Time when the
1/2	_____	temperature of
1	_____	the solid started
		to drop _____

Part II: Determination of the heat of fusion of a solid

In this part of the experiment we will essentially repeat Part I, this time measuring the temperature of the water in the calorimeter. As the liquid freezes, it will give off heat to the water. The heat given off is the heat of fusion of the liquid. We will let the sample furnish heat to the water from the time it first begins to freeze until the time it has all been converted to solid.

1. Put the test tube from Part I in the hot water bath and remelt the solid, again bringing the temperature of the liquid to about 10° above the melting point. Turn off the burner and leave the tube in the bath. Remove the thermometer, leaving as much of the liquid in the tube as possible. Wipe off the thermometer.
2. Reassemble the coffee cup calorimeter. Add exactly 150 ml of room temperature water. Measure the temperature of the water. Remove the test tube from the water bath and let it cool in air as in Part I. When the first crystals of solid appear in the tube, begin timing and put the tube in the water in the calorimeter. Gently stir the water in the calorimeter with the tube. Leave the tube in the water for the same amount of time as it took for the temperature of the solid to begin to fall in Part I. At that point the liquid should be all solidified. Remove the test tube from the water and record the maximum temperature of the water.

SAMPLE DATA TABLE   **EXPERIMENT 18 (Part II)**

volume of water   _____ ml   temp of water (initial) _____ °C

temperature change (Δt) _____ °C   temp of water (final) _____ °C

CALCULATIONS AND QUESTIONS

1. Graph the temperature-time data obtained in Part I. Show time on the x-axis, beginning with time at −2 minutes, and temperature on the y-axis. Use a full page of graph paper and choose scales which are easy to plot. Draw a continuous smooth curve through the plotted points. Sketch what the temperature-time curve would look like if you started measuring two minutes earlier than you did.
2. Briefly explain why the cooling curve you obtained in 1 has the form it does.
3. On the basis of your graph, determine the melting point of the solid you used. Select from Table 18–1 the probable identity of your unknown.
4. What effect would increasing the amount of unknown used have on the melting point?
5. Calculate the heat that went into the water as the liquid solidified in Part II. (You may assume the density of tap water is 1.00 g/cm^3 and that the specific heat is 1.00 cal/g°C.)

$$Q = \text{Heat effect} = \text{mass}_{H_2O} \times \text{Sp. Ht.}_{H_2O} \times \Delta t_{H_2O}$$

Q is the amount of heat that was released by the liquid as it solidified. Since the process is reversible, it is also the amount of heat that would be required to melt the solid. Therefore,

$$Q = \text{heat of fusion of the solid sample}$$

6. (a) The heat that is necessary to melt a solid is proportional to the amount of solid. Given the heat of fusion of the solid sample, as found in 5, calculate the heat of fusion of 1.00 g of sample:

$$\Delta H_{\text{fus (cal/g)}} = \frac{Q}{\text{mass of sample}}$$

   (b) Consult Table 18–1 again and use the heat of fusion to confirm the identity of the unknown.
7. What effect would increasing the amount of unknown used have on:
   (a) the heat effect in calories (Question 5)?
   (b) the heat effect in calories/gram (Question 6)?

EXTENSIONS

1. Sketch the predicted shape of the temperature-time curve that you would obtain with your unknown if you heated it from 10° below to 10° above its melting point. On the curve indicate where the sample is all solid, all liquid, and where it is a solid-liquid mixture.
2. Which freezes faster, hot water or cold water? Devise an experiment to be

conducted at home which answers the question. Remember to eliminate all variables except the initial temperature of the water. Report your experimental procedure and results. What is your explanation of the results?

RELATIONSHIPS BETWEEN PHYSICAL PROPERTIES AND CHEMICAL BONDING IN SOLIDS

OBJECTIVES

1. To investigate some of the physical properties of substances containing ionic, covalent, and metallic bonds.
2. To use physical properties to characterize substances as to bond type.

DISCUSSION

The type of bonding which holds the molecules, atoms, or ions together in a solid substance determines to a large degree the physical properties of that substance. Depending on the nature of the bonding, solids may be described as being ionic, molecular, metallic, or covalent network solids. Since each type of bonding affects the physical properties of a solid in different ways, one can determine the type of bonding in a solid by studying its physical properties.

Ionic solids contain ions held together by ionic bonds. These solids are typically hard and have high melting points. They are often soluble in water, but insoluble in most organic solvents. The solid does not conduct an electric current, but its aqueous solution and the pure molten material are relatively good conductors.

Molecular solids are made up of molecules. The atoms in the molecules are held together by strong covalent bonds. The forces between molecules are relatively weak. Because of the weak intermolecular forces, molecular solids tend to be soft, easily melted, and volatile. Molecular solids are likely to be insoluble in water but soluble in organic solvents. The solid does not conduct an electrical current, and neither do any of its solutions.

Metallic solids contain atoms bonded together by metallic bonds. The bonds are strong but not localized as in other substances, since the electrons in the bonds are relatively mobile. Because of the nature of the bonding, metals tend to be hard, malleable, nonvolatile, high melting, and shiny. They are very good electrical conductors in the solid or molten states. Metals are not soluble in water, unless they react with it chemically, and are not soluble in organic solvents.

Covalent network solids contain atoms held together by a network of covalent bonds that link every atom in the solid to every other atom. In such a solid the molecule is gigantic; each particle of crystal is essentially a molecule. For this reason solids of this kind are sometimes called macromolecular. Covalent network solids are hard and include diamond, the hardest of all known substances. They are nonvolatile, very high melting, and insoluble in water and organic solvents. The solid does not conduct an electric current.

Although many substances fit nicely into one of the four categories we have described, some have properties that make their classification difficult. For example, some macromolecular substances are reasonably good electrical conductors, and some molecular substances are soluble in water. So, in classifying substances by their properties you must be aware that the distinctions between the different kinds of solids are not always sharp. We may have substances in which the bonding falls in more than one category.

In this experiment we will examine the hardness, volatility, melting point, solubility, and electrical conductivity properties of several solid substances in order to establish the class of solid to which they belong.

PRELIMINARY STUDY

1. Review Sections 12.4 and 12.5 in the text.

PROCEDURE

Wear your safety goggles, apron, and gloves while performing this experiment.

1. (a) Obtain small samples (about 1 g) of sodium chloride, NaCl, paradichloro-benzene, $C_6H_4Cl_2$, silicon dioxide, SiO_2, and iron turnings. Test the hardness of each solid by putting a small amount in an evaporating dish and rubbing with the bottom of a test tube.
    (b) Test the volatility of each substance by cautiously smelling it. If you can detect an odor the substance has at least some degree of volatility.
2. In separate experiments, test each of the four substances for ease of melting. Place a pinch of the solid in an evaporating dish on a wire gauze supported on an iron ring and heat, gently at first, with a Bunsen flame. As soon as a solid melts remove the flame.

CAUTION: The vapors of any volatile substance should be assumed to be toxic. The vapors should not be inhaled directly and adequate ventilation should be used.

If a solid does not readily melt, heat it strongly for a minute or two. Some of the solids will not melt under the conditions of this experiment.

3. (a) Place a small amount (about 0.3 g) of each solid in separate test tubes, one solid to a tube. Add 5 ml of water and shake each tube. Note the relative solubilities of the solids.

   (b) Repeat the solubility test for each of the solids, using 5 ml of trichloroethane, CCl_3CH_3, as a solvent.

4. The electrical conductivity of a substance or solution may be tested with an inexpensive light bulb tester (Figure 19–1). The light bulb will light if the substance is metallic or if its solution contains ions. Test the conductivity of each solid and the liquids listed below. To test the solids, put enough solid into a 100 ml beaker to fill it to a depth of 1 cm. When testing liquids, use about 50 ml of liquid in a 100 ml beaker. The solutions may be prepared by stirring in about 5 g of the solid. (Note: The instructor may demonstrate the conductivity tests or furnish you the solutions in order to conserve chemicals.)

   (a) distilled water

   (b) water solutions of NaCl, $C_6H_4Cl_2$, SiO_2, and iron turnings

   (c) trichloroethane

   (d) trichloroethane solutions of NaCl, $C_6H_4Cl_2$, SiO_2, and iron turnings

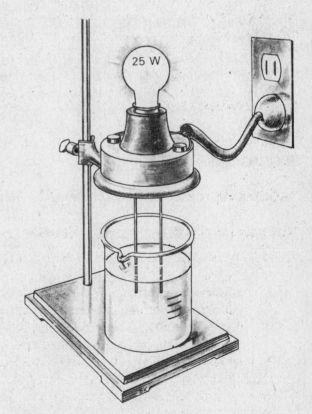

FIGURE 19-1 Testing the electrical conductivity of a solution.

5. Obtain an unknown solid and attempt to identify its bond type by repeating the tests given in Procedures 1–4.

   CAUTION: Do not carry out any tests on the unknown except as directed.

SAMPLE DATA TABLE					EXPERIMENT 19
Property Tested		Observations			
	NaCl	$C_6H_4Cl_2$	SiO_2	Fe	Unknown No. _____

QUESTIONS

1. (a) Explain the results of the hardness and volatility tests in terms of the type of chemical bond in the substance.
   (b) If a substance has no odor, does that mean that it is not volatile? Explain.
2. Arrange the four substances as best you can in order of increasing melting point. Explain the order in terms of bond type.
3. (a) Which substances are soluble in water? in trichloroethane?
   (b) Based on your solubility test results, write a general statement which relates solubility in water to the bond type of the substance being dissolved.
4. Explain the results of the electrical conductivity tests in terms of bond type.
5. Identify the bond type in each of the substances tested.
6. What is the bond type of the unknown compound tested in Procedure 5? Give the reasons for your answer.

EXTENSIONS

1. (a) Classify the following substances as ionic, molecular, or macromolecular.

   (1) I_2 (3) SO_3 (5) SiC

   (2) $CaCl_2$ (4) Na_2SO_4

   (b) Predict the relative melting points of the substances. If possible, verify your prediction by finding the actual melting points in a chemical handbook.

THE EFFECT OF TEMPERATURE ON SOLUBILITY

OBJECTIVES

1. To prepare a saturated solution with a known concentration.
2. To study the effect of temperature on solubility.
3. To determine the solubility curve for a salt.

DISCUSSION

If you add a small amount of a substance like sodium chloride, NaCl, to some water, you will find that, with stirring, all of the solid dissolves. If you continue to add NaCl, a point is finally reached at which it no longer dissolves, no matter how long you stir the mixture. The solution at that point contains all the NaCl it can hold, and is said to be saturated. At 25°C, 100 grams of water can dissolve up to 36.2 grams of NaCl, at which point the solution is saturated. The amount of solid that will dissolve in a given amount of solvent is called the solubility of the solid. The solubility of NaCl in water at 25°C is 36.2 g per 100 g of H_2O. For all solids there is a limiting amount that will dissolve in a given solvent. Some solids are very soluble in the most common solvent, water, while others are nearly completely insoluble. About 200 grams of ordinary sugar will dissolve in 100 grams of water at 25°C, but only about 2×10^{-4} grams of AgCl will dissolve under those conditions.

The solubility of a solid in a given solvent depends on the temperature of the solution. In Figure 20–1 we have plotted the solubility of NaCl in water between 0° and 100°C. The solubility increases slowly, from about 35.7 g to about 39.8 g per 100 g H_2O in that temperature interval. Usually the solubility of a solid increases with increasing temperature, sometimes very markedly. A graph like that in Figure 20–1 is called the solubility curve of the substance. Given such a graph you can find the solubility of the solid at any temperature given on the x-axis.

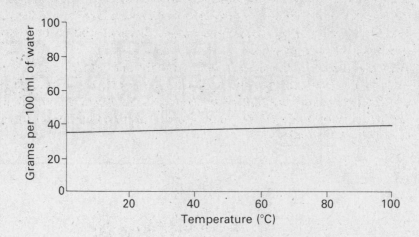

FIGURE 20–1   The solubility curve of NaCl.

In this experiment we will obtain the solubility curve for potassium dichromate, $K_2Cr_2O_7$, in water. Your part of the experiment will be to find one point on the curve. You will be assigned a certain mass of $K_2Cr_2O_7$, and your problem will be to determine the temperature at which a solution containing that mass of $K_2Cr_2O_7$ in 20 grams of water is saturated. The way you will do this will be to add the solid to the water at room temperature and stir to form a saturated solution. Not all the solid will dissolve. You will then heat the mixture in a water bath until the solid does dissolve, and ultimately it will, since the solubility of $K_2Cr_2O_7$ goes up considerably with temperature. When the sample is all dissolved, you will let the solution cool, and measure its temperature when the first crystals of $K_2Cr_2O_7$ reappear. The solution will be saturated at that point, since solid is starting to form. The solution then has a known composition, namely the one you established when you prepared the mixture. Since you know you have a saturated solution, and know the composition and temperature of that solution, you can calculate the solubility of $K_2Cr_2O_7$ in 100 g of water at that temperature. This will give you one point on the solubility curve. From similar points obtained by other students in the class, we can complete the solubility curve for $K_2Cr_2O_7$.

PRELIMINARY STUDY

1. Practice Problem: Determine the solubility of Na_2SO_4 in grams per 100 grams of water if 0.94 g of the solute in 20 g of water produces a saturated solution. (Ans: 4.7 g Na_2SO_4/100 g H_2O)

PROCEDURE

1. (a) Set up a 400 ml beaker on an iron ring and wire gauze. Add about 300 ml of tap water.

(b) Weigh out the amount of potassium dichromate, $K_2Cr_2O_7$, assigned by your instructor (4–12 g). Transfer the solid to a large test tube and add exactly 20 ml (20 g) of distilled water. Stir the mixture to dissolve as much $K_2Cr_2O_7$ as possible.

2. (a) Heat the water in the beaker with your Bunsen burner. Clamp the test tube in the water as shown in Figure 20–2. Using a thermometer, stir the mixture gently until all of the $K_2Cr_2O_7$ has dissolved. Depending on the amount of $K_2Cr_2O_7$ in your sample, the temperature required may vary from 30°C to 80°C.

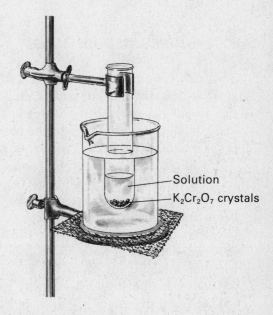

FIGURE 20–2 Apparatus for determining the solubility of $K_2Cr_2O_7$.

Solution

$K_2Cr_2O_7$ crystals

(b) When all the $K_2Cr_2O_7$ has dissolved, loosen the clamp and raise the tube out of the water bath. Turn off the burner. Reclamp the tube above and to one side of the beaker. Using a thermometer, stir the solution gently and observe it as it cools. As soon as crystallization begins, note the temperature of the solution. (It is easiest to see crystals by looking at the bottom of the tube where they will collect as soon as they begin to form.)

3. After you are certain crystallization has begun and have recorded the solution temperature, put the test tube back into the water bath and warm the solution until all of the crystals have redissolved. Repeat Procedure 2(b) to check the crystallization temperature. Your temperature readings should be within about 1 degree of each other. If necessary, rewarm the solution and repeat 2(b) again.

4. Record your data point on the blackboard. Complete your data table with the solubility data from the other students in the class.

```
┌─────────────────────────────────────────────────────────────────────┐
│  SAMPLE DATA TABLE                              EXPERIMENT 20          │
│                                                                       │
│      mass of K₂Cr₂O₇        saturation temp (°C)    g K₂Cr₂O₇/100 g H₂O │
│                                                                       │
│   1. _____ g     _____       _____    │
│                                                                       │
│   2. _____ g     _____       _____    │
│                                                                       │
└─────────────────────────────────────────────────────────────────────┘
```

CALCULATIONS AND QUESTIONS

1. Your data furnishes the amount of $K_2Cr_2O_7$ that will dissolve in 20 g of water at each temperature. For each data point calculate the amount of $K_2Cr_2O_7$ that would dissolve in 100 g of water. Your calculations will be the solubilities of $K_2Cr_2O_7$ in water at each of the temperatures reported.
2. (a) Construct the solubility curve for $K_2Cr_2O_7$. Plot solubility in grams per 100 grams of water on the y-axis and temperature on the x-axis. Extend the curve to include solubility from 10°C to 100°C.
   (b) State in words how the solubility of $K_2Cr_2O_7$ varies with temperature.
3. From your solubility curve predict
   (a) the solubility of $K_2Cr_2O_7$ at 100°C.
   (b) the solubility of $K_2Cr_2O_7$ at 10°C.
   (c) the solubility of $K_2Cr_2O_7$ at 25°C in units of moles of solid/liter of solution (molarity). The density of the solution is about 1.1 g/ml.
   (d) the mass percent of $K_2Cr_2O_7$ in its saturated solution at 50°C.
   (e) the percentage of $K_2Cr_2O_7$ that will crystallize out when its saturated solution at 100°C is cooled to 10°C.

EXTENSIONS

1. Using the solubility curves of different solutes, it is possible to develop a simple method for separating substances. Assume, for example, that you have a solution containing 90 grams of $K_2Cr_2O_7$ and 10 grams of NaCl. If we heat the solution and boil it down until it is saturated at 100°C with one of the salts, how much water will be present in the solution? (You can assume that the salts do not interact with each other.) If that solution is then cooled to 10°C, how much $K_2Cr_2O_7$ will crystallize? How much NaCl? Which salt could be recovered pure by this process? What percentage could be recovered? The process is called fractional crystallization and is used extensively to purify salts which are contaminated by small amounts of soluble species.

PREPARATION AND PROPERTIES OF SOLUTIONS

OBJECTIVES

1. To learn the procedures for making solutions of known concentration.
2. To become familiar with expressing concentration using the molarity system.
3. To determine the concentration of a solution by physical methods.
4. To determine the effect of concentration on boiling point.

DISCUSSION

Many chemical reactions are carried out in solution, since solutions offer a convenient way to make available and mix known amounts of reagents. Most substances can be dissolved in one or more solvents, so it is usually easy to find a suitable solvent in which to conduct a reaction. As you might expect, water is the most common solvent one encounters.

The Molarity of a Solution

In order to be able to measure out a known amount of reactant when it is in solution, it is necessary to know the concentration of the reactant. Concentrations can be expressed in many ways. In the previous experiment we found the concentration of saturated solutions in grams of solute per 100 grams of water. While this system of expressing concentration is very satisfactory for describing solubilities, it is not as convenient for many purposes as that involving the molarity. The molarity of the solute A is defined in the following way:

$$\text{molarity of A, } M_A = \frac{\text{moles of A in solution, } n_A}{\text{volume of solution in liters, V}} \qquad (21.1)$$

The reason molarity is useful is that it involves the volume of the solution, which is easy to measure. Knowing the molarity and volume, the amount of solute in moles follows immediately from Equation 21.1:

$$\text{moles of A} = \text{molarity of A} \times \text{volume in liters;} \qquad n_A = M_A \times V \quad (21.2)$$

To get the amount of A when concentration i expressed in other units one usually needs to weigh the solution, which is a relatively inconvenient procedure.

Preparing a Solution of Known Molarity

To make a solution of known molarity one can dissolve a known amount of solute in the solvent, and then add solvent to make the desired volume of solution. On thorough mixing, the solution will have a molarity given by Equation 21.1. Another way to prepare a solution of known molarity is to dilute a known stock solution with the proper amount of solvent. In that procedure the number of moles of solute is the same in the stock solution and the prepared solution, so Equation 21.2 can be applied to both solutions:

$$\text{moles of A} = (M_A)_1 \times V_1 = (M_A)_2 \times V_2 \qquad (21.3)$$

where 1 and 2 refer to the stock and prepared solutions respectively. To make 100 ml of a 0.5 M HCl solution from a stock 6 M HCl solution, we would find the volume of stock solution needed, by using Equation 21.3:

$$V_1 = \frac{(M_A)_2 \times V_2}{(M_A)_1} = \frac{0.5 \text{ mole/\ell} \times 0.100 \text{ \ell}}{6 \text{ moles/\ell}} = 0.008 \text{ \ell} = 8 \text{ ml}$$

The desired solution could be prepared by measuring out 8 ml of 6 M HCl and adding water with mixing until the total volume of solution was equal to 100 ml.

We can measure the volumes of solutions in several ways. The most common is probably the graduated cylinder, which you have used many times in this course. For less precise volume measurements one can use a graduated beaker or flask. When high precision is needed, we use volumetric flasks.

Physical Properties of Solutions

The physical properties of solutions differ from those of the solvent. Solutions of nonvolatile solutes boil at higher temperatures than the pure solvent, and freeze at lower temperatures. Solutions of ionic substances conduct an electric current. For a given molarity an ionic solute like NaCl will have more effect on boiling point and freezing point than a nonionic solute like sucrose because the number of solute particles per liter of solution will be larger. The boiling point elevation for 1 M NaCl will be about twice that for 1 M sucrose because there will be two moles of ions per liter with NaCl and one mole of molecules per liter with the sucrose.

Finding the Molarity of a Solution

Given a solution A, having an unknown molarity, one can determine the molarity of A by various means. The simplest one, at least in principle, would be to take a known volume of solution, boil off the solvent, and weigh the dry solute. If the solution is colored, one can compare its color to that of a solution of known concentration of the same solute. Two solutions which have the same intensity of color will have the same concentration. Another way to match intensity would be to look down two test tubes, one containing an unknown and the other a known solution (Figure 21–1). If you adjust the volume of the unknown until the color intensity as seen down the tube matches that of a volume of the known solution, you will be looking through the same total amount of solute, so Equation 21.3 can be applied to the two solutions:

$$(M_A)_1 \times V_1 = (M_A)_2 \times V_2 \qquad\qquad (21.3)$$

where 1 and 2 refer to the known and unknown solutions respectively. To find the molarity of the unknown we would measure the volumes of the known and unknown solutions that match in color intensity. On substituting in Equation 21.3, we obtain:

$$(M_A)_2 = \frac{V_1}{V_2} \times (M_A)_1 \qquad\qquad (21.4)$$

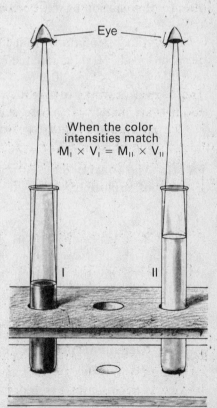

FIGURE 21-1 Determining the molarity of a solution by color matching.

In this experiment we will carry out all of the procedures we have described here. We will make solutions of known molarity by dilution of a stock solution and by dissolving a known mass of solute in enough solvent to make a measured volume of solution. We will measure the boiling points of these solutions. We will find the concentration of a solution of NaCl by evaporation of the solvent and the concentration of an unknown violet solution of $KMnO_4$ by matching the color intensity of the unknown with that of a known.

PRELIMINARY STUDY

1. Review Chapter 13 in the text.
2. The results of this problem are needed in Part I of the experiment. Complete the calculation before coming to the laboratory.
   (a) Calculate the mass of sucrose, $C_{12}H_{22}O_{11}$, needed to make 25.0 ml of a 2.00 M solution.
   (b) Calculate the volume of a 3.00 M $CaCl_2$ solution needed to make 25.0 ml of a 2.00 M solution by dilution with water.

PROCEDURE

Wear your safety goggles, apron, and gloves while performing this experiment.

Part I: Preparation of solutions of known concentrations.

Use your calculations from the PRELIMINARY STUDY to make the following solutions in 50 ml graduated beakers.
   (a) 25 ml of a 2.00 M $C_{12}H_{22}O_{11}$ solution beginning with pure sucrose. Don't forget to mix the solution well.
   (b) 25 ml of a 2.00 M $CaCl_2$ solution beginning with a 3.00 M stock solution. Use a graduated cylinder for volume measurements and mix well.

Part II: The effect of concentration on boiling point of a solution

Heat the two solutions you prepared in Part I to the boiling point. Use an iron ring and wire gauze to support the 50 ml beakers. When the solution is boiling gently, put the thermometer in the liquid until the bulb is completely immersed. Read and record the temperature after it has become steady. When you have completed these measurements, determine the boiling point of water by boiling a 25 ml sample in a 50 ml beaker.

Part III: Determining the molarity of solutions of NaCl and $KMnO_4$

1. (a) Pour about 25 ml of water into a 50 ml beaker. Add sodium chloride,

NaCl, until, after stirring for at least a minute, some NaCl remains undissolved.

(b) Weigh a dry 100 ml beaker with a watch glass that will cover it. Filter the mixture in 1(a), and pour 10 ml of the filtrate, as measured in a graduated cylinder, into the beaker. Boil the solution to remove the water. When the liquid begins to bump, cover the beaker with the watch glass to keep material from spattering out. Continue to heat, gently, until the water is all evaporated from the solid. By judicious use of the burner, drive off any water that condenses on the watch glass or the walls of the beaker. Remove the watch glass, and put it upside down on the lab bench without disturbing any solid that has stuck to it. Let the beaker cool down to room temperature, re-cover with the watch glass and weigh the beaker, watch glass and NaCl.

2. (a) Your teacher has prepared some solutions of $KMnO_4$ of known molarity. These solutions have been arranged on a rack in order of increasing concentration. Obtain a solution of $KMnO_4$ of unknown concentration in a test tube like the one on the rack and compare its color with those of the known solutions. Record the molarity of the known solution whose color is nearest to, and a little deeper than, that of your unknown.

(b) Prepare 20 ml of a solution of $KMnO_4$ having the same concentration as that of the solution you selected in 2(a). Make this solution by proper dilution of the stock 1.0×10^{-3} M $KMnO_4$ solution, mixing well.

(c) Pour 20 ml of your unknown solution, as measured in a graduated cylinder, into a regular test tube or flat-bottom vial. To another container of the same size, add the solution you prepared in 2(b) until the color intensity of the unknown matches that of the known when both solutions are viewed down the tubes against a well-illuminated white background (Figure 21–1). Use a medicine dropper to adjust the final volume. In a graduated cylinder, measure the volume of prepared solution required to produce the matching color intensity.

SAMPLE DATA TABLE   **EXPERIMENT 21**

Part I

mass of $C_{12}H_{22}O_{11}$ _____ g   vol. of 3.00 M $CaCl_2$ soln. _____ ml

Part II

BP of H_2O _____ °C   BP of $C_{12}H_{22}O_{11}$ soln. _____ °C

BP of $CaCl_2$ soln. _____ °C

Part III

mass of beaker and watch glass _____ g   vol. of NaCl soln.   _____ ml

mass of beaker, watch glass   mass of NaCl   _____ g

and NaCl   _____ g

$KMnO_4$ unknown no.   _____   molarity of known

volume of unknown solution   solution selected   _____ M

used for color matching   _____ ml   volume of prepared solution

volume of stock solution used   needed for color matching _____ ml

in Procedure 2(b)   _____ ml

CALCULATIONS AND QUESTIONS

1. (a) Determine the boiling point elevation (BP soln. – BP H_2O) of each solution as measured in Part II. Calculate the ratio:

$$\frac{\text{boiling point elevation of } CaCl_2 \text{ solution}}{\text{boiling point elevation of } C_{12}H_{22}O_{11} \text{ solution}}$$

   (b) Explain as best you can the boiling points you observed.
2. (a) Using the results obtained in Part III, Procedure 1, calculate the molarity of the saturated sodium chloride solution you prepared (Eqn. 21.1).
   (b) If the density of the saturated solution is 1.2 g/cm^3, what is the mass percent of NaCl in the solution?
3. Using the data obtained in Part III Procedure 2, find the molarity of your unknown $KMnO_4$ solution (Eqn. 21.4).

EXTENSIONS

In some schools colorimeters are available which measure color intensity automatically. With such instruments one can find the percentage of the light of any given color that is transmitted by a solution. If your school has such an instrument, use it to find the concentration of your unknown sample in Part III, Procedure 2. Your teacher will describe the use of the instrument. (Section 21.3 in your text has some information on colorimeters.) Compare the results you obtain with the colorimeter with those you found by color matching and with the actual value as furnished by your teacher.

ISOMERISM IN ORGANIC CHEMISTRY

OBJECTIVES

1. To better understand structural and geometric isomerism.
2. To build molecular models of hydrocarbon isomers.
3. To practice writing structural formulas.

DISCUSSION

As you know, in inorganic chemistry we can usually identify a substance by its simplest formula. There is only one substance having the formula NaOH and only one with the formula $CuSO_4$. If we find a compound in which the Na:O:H atom ratio is 1:1:1, we know that that compound must be sodium hydroxide.

In organic chemistry the situation is somewhat different. There we deal with molecular formulas, which give more information than simplest formulas since they tell us the actual number of each kind of atom present in the molecule. The molecular formula $C_3H_6Cl_2$ indicates that each molecule of the substance with that formula contains 3 carbon, 6 hydrogen, and 2 chlorine atoms. Rather surprisingly, the formula does not represent just one substance, but several, all of which have the same molecular formula. The reason is that the atoms in the molecules with this formula can be attached to one another in more than one way, each of which produces a different substance. The molecular formula $C_3H_6Cl_2$ is associated with each of the molecules whose structural formulas are shown below:

$$
\begin{array}{ccc}
\text{(I)} &
\begin{array}{ccc}
\text{H} & \text{H} & \text{H} \\
| & | & | \\
\text{Cl} - \text{C} - \text{C} - \text{C} - \text{Cl} \\
| & | & | \\
\text{H} & \text{H} & \text{H}
\end{array}
&
\text{(II)} \quad
\begin{array}{ccc}
\text{H} & \text{H} & \text{H} \\
| & | & | \\
\text{Cl} - \text{C} - \text{C} - \text{C} - \text{H} \\
| & | & | \\
\text{Cl} & \text{H} & \text{H}
\end{array}
\end{array}
$$

(III)   H   H   H   (IV)   H   Cl   H
     | | | | | |

Cl – C – C – C – H   H – C – C – C – H

     | | | | | |

     H Cl H   H Cl H

Molecules like those we have drawn, all of which have the same formula, are called structural isomers. There are four structural isomers having the molecular formula $C_3H_6Cl_2$, each with its own characteristic properties. Isomerism of this sort is common in organic chemistry, even with fairly simple molecules, and is responsible in large measure for the very large number of known organic substances.

Note that in each of the isomeric structures we have shown, the C – Cl bonds are arranged differently in the molecule. It is not possible to convert one isomer to another by rotation of the molecule or by rotation around carbon-carbon bonds. If such conversion were possible, we would not have isomers, but simply different views of the same molecule.

There are several other possible types of isomerism in organic chemistry. The only other kind we will encounter in this experiment is called geometric isomerism. This occurs in molecules containing carbon-carbon double bonds, and is due to the fact that there is no rotation around such bonds. Around carbon-carbon single bonds we have free rotation (actually at a frequency much faster than that of any propeller that man has ever built), but around carbon-carbon double or triple bonds no such rotation occurs. This means that the following two structural formulas represent different substances:

Cl\   /Cl   H\   /Cl

    C = C   C = C

H/   \H   Cl/   \H

        cis   trans

The isomer on the left is called cis and that on the right is called trans, indicating that the two Cl atoms are on the same side, and on opposite sides of the molecule, respectively. We have isomers (same formula), with the same bonding (same structure), but different geometry, so we have two geometric isomers with this formula.

In this experiment you will make models for some organic molecules that may exhibit structural or geometric isomerism. The model kits we will use are the same as in Experiment 17. All organic compounds obey the octet rule, so the same principles will apply as in the earlier experiment. You will find that some molecular formulas lead to isomers, and that the models allow you to identify the different isomers much more easily than do structural drawings in a plane. For each model you build, you should draw the structural formula that goes with it.

PRELIMINARY STUDY

1. Review Chapter 14.
2. Find the structural isomers having the molecular formula C_4H_9Cl. Draw the structural formula for each isomer.

PROCEDURE

1. Obtain a molecular model kit. Use four-hole balls for carbon atoms, 1-hole balls for hydrogen atoms, and 1-hole balls of a different color for chlorine atoms. Use sticks for single bonds and springs for double bonds.
2. Build models for each molecule listed. All the molecules obey the octet rule. This means that in the models, all of the holes in all of the balls must be filled when the model is complete. (Chlorine atoms will have eight electrons, one pair forming the bond to carbon and the other three pairs unshared. In the model the three unshared pairs are not shown, and can be considered to be part of the ball representing the chlorine atom.) Some of the molecular formulas can be associated with two or more isomers. Isomers may arise by branching the main carbon-carbon chain, by moving a carbon-carbon multiple bond, and by moving chlorine atoms. Where double bonds are present, be alert to the possibility of geometric isomers. In your data table, for each formula draw the structural formula of all the possible isomers.

(a)	CH_2Cl_2	(g)	C_4H_{10}
(b)	$C_2H_4Cl_2$	(h)	C_4H_8
(c)	C_3H_8	(i)	C_5H_{12}
(d)	C_3H_7Cl	(j)	C_6H_6 (aromatic only)
(e)	C_2H_4	(k)	$C_6H_5(CH_3)$ (aromatic only)
(f)	$C_2H_2Cl_2$	(l)	$C_6H_4Cl_2$ (aromatic only)

SAMPLE DATA TABLE   **EXPERIMENT 22**

Compound	Structural Formulas

QUESTIONS

1. Which of the substances with the molecular formulas given in Procedure 2
   (a) would be expected to exist in only one isomeric form?
   (b) would have geometric isomers?
   (c) are unsaturated?
   (d) are planar?
2. On the basis of the models you built, what are the C-C-C bond angles in C_3H_8, C_4H_8, and C_6H_6?

EXTENSIONS

1. A substance with the molecular formula C_4H_8 can exist in a cyclic form as well as a linear one. Make the model for the cyclic form of the molecule, using springs for the C-C bonds. Given the fact that C-C-C bonds are usually at the tetrahedral angle, what do you think the geometry of the C_4H_8 cyclic molecule would be?

PROPERTIES OF HYDROCARBONS

OBJECTIVES

1. To prepare two hydrocarbon gases.
2. To carry out some simple characteristic reactions of four groups of hydrocarbons.

DISCUSSION

Organic compounds which contain only carbon and hydrogen belong to that family we call hydrocarbons. There are several classes of hydrocarbons, which are distinguished by the kinds of carbon-carbon bonds present in their molecules. Molecules in which all carbon-carbon bonds are single bonds are called alkanes and are said to be saturated. Molecules with double bonds are alkenes, those with triple bonds are alkynes. The alkenes and alkynes are both classified as unsaturated hydrocarbons. Alkanes, alkenes, and alkynes may exist in both linear and cyclic forms. Those hydrocarbons with six membered rings and resonance forms in which single and double carbon-carbon bonds alternate have rather distinctive properties and are classified as aromatic.

The physical properties of the hydrocarbons, such as boiling point and solubility, depend mainly on their molecular weight and not on the kinds of carbon-carbon bonds present. Hydrocarbons are soluble in other hydrocarbons (like dissolves like), and also usually dissolve in any nonpolar organic solvents; hydrocarbons are not appreciably soluble in water. Hydrocarbons of molecular mass less than about 100 and greater than about 70 are usually liquids at 25°C; their odors are characteristic, and do seem to depend to some extent on the kinds of carbon-carbon bonds present.

All hydrocarbons burn in air, and in excess air form carbon dioxide and water as products. Natural gas, fuel oil, and gasoline are all hydrocarbons, and are our most important fuels. Except for combustion, the alkanes undergo relatively few reactions. The alkenes and alkynes are much more reactive chemically. A typical reaction occurs with elementary halogens, adding halogen atoms to carbon atoms connected by double or triple bonds:

$$
\begin{array}{cc}
H & H \\
| & | \\
H-C=C-H + Cl-Cl \rightarrow H-C-C-H \\
\end{array}
$$

$$
\begin{array}{cc}
H & H \\
| & | \\
H-C-C-H \\
| & | \\
Cl & Cl
\end{array}
$$

This is called an addition reaction. With alkynes, an alkene is formed if one mole of halogen is added, and an alkane if two moles are added. Aromatic compounds do not add halogens under ordinary conditions.

Hydrocarbons can be made by many reactions, and are usually prepared from other hydrocarbons. The petrochemical industry is based on processes in which one hydrocarbon is converted to another. In the laboratory, two simple hydrocarbons can be made by the following reactions:

$$NaC_2H_3O_2(s) + NaOH(s) \rightarrow CH_4(g) + Na_2CO_3(s) \qquad (23.1)$$

$$CaC_2(s) + 2\,H_2O \rightarrow C_2H_2(g) + Ca(OH)_2(s) \qquad (23.2)$$

In this experiment we will prepare two gaseous hydrocarbons by reactions 23.1 and 23.2 and will observe their behavior on combustion and on treatment with Br_2. We will examine the solubility properties and reactions with Br_2 of the set of compounds listed below:

cyclohexane C_6H_{12} (alkane)   cyclohexene C_6H_{10} (alkene)   toluene $C_6H_5CH_3$ (aromatic)   naphthalene $C_{10}H_8$ (aromatic)

Although these substances appear to have similar molecular structures, their chemical behaviors are quite different since they belong to different classes of hydrocarbons.

PRELIMINARY STUDY

1. Review Chapter 14.
2. Given a sample of a liquid hydrocarbon, along with its molecular formula, C_5H_{10}, how could you determine, using both experiments and theory, whether it is an alkane, an alkene, an alkyne, or an aromatic compound?

PROCEDURE

Wear your safety goggles, apron, and gloves while performing this experiment.

Part I: Preparation and properties of methane, CH₄.

1. Mix 4 g of anhydrous sodium acetate, $NaC_2H_3O_2$, with 8 g of "soda lime" (a mixture of CaO and NaOH) in a mortar. Put the mixture in a large test tube with a stopper that has a jet-tip delivery tube with a right-angle bend inserted through the stopper (Figure 23–1).

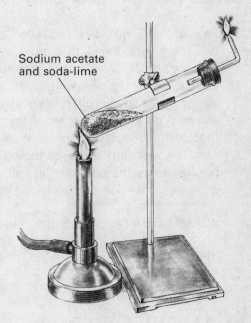

Sodium acetate
and soda-lime

FIGURE 23–1 Apparatus for producing and burning methane gas.

2. Heat the mixture, which will quickly begin to evolve methane. Invert a small test tube over the jet, collect some of the gas (which is lighter than air), and ignite the gas in the tube with your Bunsen burner. Using the burning gas in the tube, ignite the gas coming out of the jet. Do not ignite the gas from the jet with your burner, since you might get an explosion. When the CH₄ coming from the jet is pure, it will burn long enough in the test tube for you to use the flame to light the jet. Move the burner away from the tube and extinguish the jet flame.

3. Fill a small test tube about one-third full with water. Add 2–3 drops of bromine water. Turn the delivery tube so that it is pointing down, and again heat the mixture to produce methane gas. Bubble the gas through the bromine water and watch carefully to see if a reaction takes place.

Part II: Preparation and properties of acetylene, C₂H₂.

1. Drop a small piece of calcium carbide, CaC_2, into a 400 ml beaker 2/3 full of

tap water. The gas evolved is acetylene, C_2H_2. Collect two small test tubes of gas by filling the tubes with water and holding them, inverted, over the carbide as it gives off the gas. Cork the tubes under water leaving about 1 cm of water in each tube. Store the tubes upside down. Test the water in the beaker for OH^- ions by adding 2 drops of phenolphthalein.

2. Test the combustibility of the gas by uncorking one of the tubes near a flame. Compare the acetylene flame to that produced by methane.
3. Uncork the second tube of gas and quickly add 2–3 drops of bromine water. Cork the tube, shake, and note any reaction.

Part III: Properties of some cyclic hydrocarbons.

1. Pour about 2 ml of cyclohexane, cyclohexene, and toluene into three small test tubes, one liquid to a tube. Add 3–4 drops of bromine water to each liquid. Cork the tubes, shake, and note any reaction.
2. Obtain about 1 g of naphthalene and note its odor. Divide the naphthalene among three small test tubes and test its solubility in different solvents by adding 3 ml portions of cyclohexane, toluene, and water, one solvent to each tube. Stopper each tube and shake well.
3. Test the solubility of cyclohexane in water and toluene. Use about 2 ml portions of each liquid, mixing well.

SAMPLE DATA TABLE   **EXPERIMENT 23**

Procedure   Observations

QUESTIONS AND CALCULATIONS

1. (a) Which compounds reacted with bromine in Parts I–III? Assuming that one mole of Br_2 added to one mole of compound, write the structural formulas of the products formed.
   (b) Which classes of organic compounds are represented by the compounds which reacted with bromine?
2. Write a balanced chemical equation for the following reactions:
   (a) the combustion of acetylene in air.
   (b) the reaction of bromine with one of the compounds used in this experiment.
3. Given a sample of gas which might be either methane or acetylene, describe two experiments you might do to identify the gas.
4. What experimental evidence do you have which supports the validity of Equation 23.2?

5. How much calcium carbide would it take, by Equation 23.2, to make 100 liters of acetylene at STP?

6. (a) In which solvent is naphthalene most soluble? least soluble? Explain these observations as best you can.

   (b) Comment on the relative solubilities of cyclohexane in toluene and water.

   (c) Suggest a liquid in which cyclohexene would be likely to dissolve readily.

EXTENSIONS

1. Natural gas consists of a mixture of methane, CH_4, and ethane, C_2H_6. Suggest an experimental method by which you might determine the fraction of the molecules in the mixture that are methane. This is also the mole fraction of methane. Obtain your teacher's approval of your method. Carry out the experiment and make the necessary calculations.

ORGANIC FUNCTIONAL GROUPS

OBJECTIVES

1. To become familiar with the important organic functional groups by building molecular models.
2. To illustrate organic reactions with models.
3. To learn the names of some common organic compounds.

DISCUSSION

In a previous experiment hydrocarbons were classified as saturated, unsaturated, or aromatic. Most organic compounds contain elements in addition to carbon and hydrogen. The most common such element is oxygen. Organic compounds which contain oxygen are best classified in terms of their functional group. The functional group, which depends on the bonding arrangement on the oxygen atom, imparts particular properties to the molecule. It is convenient to consider organic compounds with the same functional group as belonging to the same class. For example, organic compounds containing the hydroxyl group, —OH, are classified as alcohols.

The geometrical arrangements of the functional groups in organic compounds are best visualized with 3-dimensional molecular models. In this experiment you will construct models of molecules which contain functional groups involving oxygen. In addition, you will use models to demonstrate some of the reactions of compounds which have these groups. The models will aid you in writing equations to describe the reactions.

PRELIMINARY STUDY

1. Review Chapter 15 in the text. All of the compounds and nearly all of the reactions in the experiment are discussed in the text. Complete Procedures 1(a) and 1(b) before coming to the laboratory.

PROCEDURE

1. (a) List the following classes in the first column of your data table. (Leave about a 3 cm space between each formula.)

alcohol	ketone	ether
aldehyde	acid	ester

   (b) Complete the second column by drawing the structural formula for a 3-carbon molecule of each class. Name the compounds whose molecules you have drawn. Consult the text for any names you don't recall.

   (c) A ball and stick molecular model kit will be used as in previous experiments. Use the 4-hole balls to represent carbon atoms and single-hole balls to represent hydrogen atoms. Oxygen atoms obey the octet rule in organic compounds. Of the four pairs of electrons around oxygen, two are unshared. In our models we will not include the unshared pairs, so oxygen atoms will be represented by balls with two holes. These holes are drilled at an angle of about $110°$ with respect to each other. The balls for oxygen are of a different color from the others, to give added emphasis to the location and geometry of the functional groups. Use sticks for single bonds and springs for multiple bonds.

   (d) Construct a molecular model for each structure you drew in (b). Since organic molecules obey the octet rule, all the holes in all the balls should be filled in each completed model. Compare the model with the structural formula on which it is based. Note the arrangement of the atoms in each functional group and describe its geometry in words.

2. In this section you will use molecular models to represent the reactants and products of organic reactions. For the reactions given below, build the models of the reactants and convert the models to models of the products by rearranging the atoms. Draw the structural formula of each organic product in your data table. Identify its class and name the compound.

   (a) Oxidation of an alcohol to an aldehyde:

   $$methyl\ alcohol + O_2 \rightarrow product + water$$

   (b) Condensation of two alcohol molecules:

   $$ethyl\ alcohol + ethyl\ alcohol \rightarrow product + water$$

   (c) Oxidation of an aldehyde to an acid:

   $$propionaldehyde + O_2 \rightarrow product$$

   (d) Condensation of an alcohol and an acid:

   $$methyl\ alcohol + acetic\ acid \rightarrow product + water$$

```
SAMPLE DATA TABLE                         EXPERIMENT 24

Procedure 1:

               Structural            Compound          Geometry of the
   Class       Formula               Name              Functional Group



Procedure 2:

               Structural Formula    Class of          Name of
   Reaction    of Product            Product           Product
```

QUESTIONS

1. (a) Which of the molecules in Procedure 1(b) can form intermolecular hydrogen bonds? To what classes do these molecules belong? Using structural formulas, show one example of hydrogen bonding between molecules.

   (b) Which of the molecules in Procedure 1(b) can have structural isomers in the same class? (For example, is there more than one kind of propyl alcohol?) Draw the structural formulas of all such isomers.

   (c) Which molecules in Procedure 1(b) are isomeric with each other? Hint: Check the molecular formula of each compound.

2. Using structural formulas write a balanced chemical equation for each reaction of Procedure 2.

3. Identify the class of each of the following compounds; name each compound as best you can:

   (a) $HCHO$ (d) $CH_3OCH_2CH_2CH_3$
   (b) CH_3COOH (e) $CH_3CH_2CHOHCH_2CH_3$
   (c) $CH_3COCH_2CH_3$ (f) $CH_3COOCH_2CH_3$

EXTENSIONS

1. Methyl benzoate is an ester used in some perfumes. Build a model of this ester by reacting a molecule of benzoic acid (C_6H_5COOH) with a molecule of methyl alcohol. Write the balanced equation for the reaction using structural formulas.

ORGANIC SYNTHESES

OBJECTIVES

1. To carry out two organic reactions of industrial and historical significance.
2. To study the properties of a soap and of a coal tar dye.

DISCUSSION

Soap

Some organic reactions were used to make products of practical importance before the actual reactions were understood. For centuries people have boiled animal fat with the water extract from wood ashes to make the substance we call soap. The product was useful for cleaning clothes and, occasionally, even taking a bath. Soap was appreciated, if not understood. Today we know that fats are esters of the trihydroxy alcohol called glycerol and long chain carboxylic acids, such as stearic acid, $CH_3(CH_2)_{16}COOH$. In the presence of hydroxide ions in hot solution, the fat undergoes the following reaction, called a saponification:

$$
\begin{array}{c}
\underset{H}{\overset{H}{|}} \quad \underset{}{\overset{O}{\|}} \\
H-C-O-C-R \\
| \quad\quad O \\
| \quad\quad \| \\
H-C-O-C-R + 3Na^+(aq) + 3OH^-(aq) \rightarrow \\
| \quad\quad O \\
| \quad\quad \| \\
H-C-O-C-R \\
| \\
H
\end{array}
\quad
\begin{array}{c}
H \\
| \\
H-C-OH \quad\quad\quad O \\
| \quad\quad\quad\quad\quad \| \\
H-C-OH + 3\,[Na^+ + {}^-O-C-R] \\
| \\
H-C-OH \\
| \\
H
\end{array}
\tag{25.1}
$$

animal fat + sodium hydroxide → glycerol + soap
                        solution

The soap which is formed is moderately soluble in water, and is a sodium salt of a fatty acid. In the old process, the hydroxide ions came from the ashes, but today we use a solution of sodium hydroxide. In the process we will employ to make soap in this experiment, we will add some ethyl alcohol to the fat-NaOH mixture in order to increase the solubility of the fat in the aqueous layer, and so speed up the saponification.

Soaps get their cleaning power from the minus ion of the acid, $R - \overset{\overset{\displaystyle O}{\displaystyle \|}}{C} - O^-$, which combines the water solubility of the polar $- \overset{\overset{\displaystyle O}{\displaystyle \|}}{C} - O^-$ group with the fat dissolving property of the long nonpolar CH_2 chain in the R group. Mixing a greasy material with a soap solution produces an emulsion containing the grease dispersed in the solution. The emulsion can be washed away with water and the grease thereby removed.

Soap has the disadvantage that its solubility is very low in solutions containing Ca^{2+} or other plus ions found in "hard water". For this reason soaps have been replaced for many purposes with detergents, which are similar in structure to soaps but have much higher solubilities in hard water.

$$CH_3 - (CH_2)_{16} - \overset{\overset{\displaystyle O}{\displaystyle \|}}{C} - O^-, Na^+ \qquad CH_3 - (CH_2)_{16} - \text{⟨benzene ring⟩} - \overset{\overset{\displaystyle O}{\displaystyle |}}{\underset{\underset{\displaystyle O}{\displaystyle |}}{S}} - O^-, Na^+$$

a soap   a detergent

Coal tar dyes

Some very important discoveries in chemistry were made by accident. The second part of this experiment involves such a discovery. Coal tar, a black viscous liquid obtained by heating soft coal in the absence of air, contains many aromatic compounds including naphthalene, and aniline:

naphthalene   aniline

Compounds such as these were of great interest to chemists in the mid-nineteenth century, and a great deal of research was done on them. In 1853 William Perkin, 15 years old at the time, went to the Royal College in London to study chemistry under a professor named Hofmann, who was working with coal tar chemicals. When he was 18 Perkin tried to synthesize quinine, a chemical used to treat malaria, by reaction of a rather complex coal tar chemical with potassium dichromate. The reaction failed, but then Perkin, for no good reason except his interest, treated a simpler related compound, aniline, with potassium dichromate. He obtained a black precipitate from which he extracted, by heating with water, a red-purple substance. He found that the substance made a good dye, which he called mauve. Perkin quit his job at the College and went into the dye business. Material dyed with mauve became so popular in England and France that the following decade was called the mauve decade. Perkin became wealthy from his product but continued his scientific work. At the age of 36 he retired from business and became a full-time research chemist.

Perkin's discovery involved luck combined with curiosity and imagination. The net result was not only a lot of mauve clothing but also the beginning of synthetic coal tar chemistry. Many new synthetic coal tar dyes were discovered in the years immediately following the success of mauve, and, in a very real sense, the chemical industry was born. Clearly, science moves ahead in unpredictable ways, many of which include luck as well as thought.

In this experiment we will repeat Perkin's synthesis of mauve. We will then use the dye on a piece of cloth so you can see what it looks like. Mauve may not be your favorite color, but in the days when it was the rage the alternatives all were derived from plants, and they had been around for a long time.

PRELIMINARY STUDY

1. Review Section 15.7 in the text.
2. Distinguish between a fat and a fatty acid. Draw the structural formula of a fatty acid with 16 carbon atoms.

PROCEDURE

Wear your safety goggles, apron, and gloves while performing this experiment. In Part II, be careful to avoid contact with aniline. If you should get some on your skin or clothing, wash it off immediately with soap and water.

Part I: Soaps and detergents

1. (a) Weigh out 10 g of lard or cottonseed oil and place in a 250 ml beaker. Add 10 ml of ethyl alcohol and 15 ml of 6 M NaOH.
   (b) Place the beaker on a wire gauze supported on a ring stand. Heat the mixture gently while stirring.

   CAUTION: Keep the flame away from the top of the beaker to prevent the alcohol from burning as it slowly evaporates.

   Continue to heat the mixture, without boiling it. The alcohol will gradually evaporate, finally leaving a pasty product. When the odor of alcohol above the beaker is no longer evident, put the beaker on the lab bench and let it cool.
   (c) Add 50 ml of a saturated NaCl solution to the pasty material and stir thoroughly. This process is called "salting out" and will precipitate the soap. Pour off the liquid, holding back the soap with a stirring rod. Rinse the soap with icewater, and press the soap into a cake.

2. (a) Take a small amount of soap and try to wash your hands with it. How does it feel and how does it work? Record your observations.

(b) Put 10 drops of kerosene (or household oil) in a test tube with 10 ml of water and shake. An emulsion or suspension of tiny oil droplets in water will be formed. Let it stand for a few minutes and observe again. While waiting, prepare another test tube as before but also add a small amount of your soap (about 1/2 g). Shake this tube thoroughly and compare the relative stabilities (resistance to separating into two phases) of the two emulsions. Repeat the above test with a synthetic detergent.

(c) Take about 1 g of your soap and add it to 30 ml of distilled water in a 100 ml beaker. Dissolve the soap by warming the solution gently while stirring. Pour equal volumes of the soap solution into three test tubes. Test the solutions separately with 5 drops of 0.1 M solutions of $CaCl_2$, $MgCl_2$, and $FeCl_3$. Note the results. Repeat the above tests with a synthetic detergent solution.

(d) Dissolve another gram of your soap in 20 ml of water. Test the solution with a few drops of phenolphthalein. Add dilute sulfuric acid, 3 M H_2SO_4, dropwise until the solution is acidic (colorless in phenolphthalein). The reaction produces the fatty acid of the fat which was used to make the soap. Are fatty acids soluble in water? Repeat the above tests with a synthetic detergent solution.

Part II: Mauve, a synthetic coal tar dye

1. Your teacher will add 5 ml of impure aniline into your 50 ml beaker. Add dilute sulfuric acid, 3 M H_2SO_4, drop by drop, stirring well after each drop. You will form a solid salt, aniline sulfate. Add acid until all of the aniline is converted to a light-colored soft solid. Then add a few more drops of the acid.

2. Add 0.5 g of potassium dichromate, $K_2Cr_2O_7$, and stir well. The solid will turn black and become shiny. (The reaction is complex. Mauve is only one of the products formed.)

3. Wash the black precipitate 3 or 4 times with 10 ml water, stirring well each time and decanting the washings. This will remove the excess potassium dichromate and sulfuric acid from the mixture.

4. Add about 40 ml of water to the black solid in the beaker. Heat on a wire gauze/iron ring over a Bunsen burner, with stirring, until the precipitate is dispersed and the liquid is near the boiling point. Let cool for a few minutes.

5. Filter the mixture through filter paper in a funnel, collecting the filtrate in a clean beaker. Discard the filter paper and the black residue it contains. The liquid filtrate should be reddish-violet in color. This is mauve dye in solution.

6. Put a small piece of white cotton cloth into a clean, empty 100 ml beaker. Add enough tannic acid solution to cover the cloth. Heat the beaker on a wire gauze/iron ring with a bunsen burner. Boil the cloth in the tannic acid solution for a few minutes. The tannic acid prepares the cloth to accept the dye. (It is called a mordant and binds the dye to the cloth.)

7. Remove the beaker from the heat. Using tongs, take the cloth out of the acid solution and rinse it in cold water.
8. Place the damp cloth in the beaker containing the filtrate from Procedure 5. Put the beaker on the wire gauze/iron ring and boil the fabric in the dye for a minute or two. Using tongs, remove the cloth from the beaker and rinse with tap water. The cloth should have a dusky purple color, from which the mauve decade took its name.

CALCULATIONS AND QUESTIONS

1. If the fat in Equation 25.1 is derived from stearic acid, what is R?
2. The predominant fatty acid in lard is oleic acid, $C_{17}H_{33}COOH$. Is this acid saturated or unsaturated?
   (a) Draw the structural formula of the fat derived from oleic acid, glycerol trioleate.
   (b) Write the equation for the making of a soap from glycerol trioleate.
3. Discuss the results of the soap tests carried out in Procedure 2 of Part I. Contrast the soap results with those obtained with a synthetic detergent. What advantages do detergents have over soaps?
4. If a mole of fat weighs about 900 grams and a mole of NaOH weighs 40 grams, which reagent was in excess in the reaction you carried out in Procedure 1 of Part I? Does your answer seem reasonable? Why?
5. Perkin was trying to make quinine, $C_{20}H_{24}N_2O_2$, from allyltoluidine, $C_{10}H_{13}N$ when he discovered mauve. The reaction he had in mind was:

$$2 C_{10}H_{13}N + 3 [O] \rightarrow C_{20}H_{24}N_2O_2 + H_2O$$

where the oxygen was furnished by potassium dichromate. Why was his planned synthesis very likely to fail?
6. Mauve is a very complex substance, as many dyes are. What are some properties that a good dye must have? Why is having the right color not sufficient to make a substance a good dye?

EXTENSIONS

1. Consult an encyclopedia, a text on the history of chemistry or technology, or an organic chemistry text to find out more about dyes. Draw the structural formulas of a dye obtained from plants, an azo dye, and Perkin's mauve. What features do these structures have in common?
2. Prepare fluorescein, another coal tar dye.
   (a) Add 0.2 g phthalic anhydride and 0.5 g resorcinol to a regular test tube.
   (b) Carefully add about 6 drops concentrated (18M) H_2SO_4 to the tube. Warm the contents very gently with a burner until a deep color appears.

CAUTION: Be careful in handling concentrated H_2SO_4. Wash off any spilled acid immediately with plenty of water.

(c) After the reaction mixture has cooled, add 10 ml 2 M NaOH, and stir. Pour the dye into a small beaker of water and observe its color and fluorescence.

PREPARATION AND PROPERTIES OF POLYMERS

OBJECTIVES

1. To prepare three polymers which have commercial importance.
2. To become familiar with some physical properties of polymers.
3. To practice writing the structural formulas of monomers and polymers.

DISCUSSION

Although most organic molecules are reasonably small, consisting of perhaps as many as 20 atoms or less, there is one group of organic compounds in which the molecules are very large, with as many as a thousand, or even ten thousand, atoms in one molecule. Such molecules are called polymers, since they contain many (poly) similar units linked together. The units are called monomers. Most polymers consist of long chains of monomer units which are bonded chemically end to end.

Polymers fall into two classes, called addition polymers and condensation polymers. In addition polymers, the monomers always contain at least one carbon-carbon double bond, which disappears when the polymer forms; a typical equation for addition polymerization is:

$$n \begin{bmatrix} H \\ \diagdown \\ C = C \\ \diagup \quad \diagdown \\ H \qquad H \end{bmatrix} \rightarrow \begin{array}{cccc} H & R & H & R \\ | & | & | & | \\ -C-C-C-C- \\ | & | & | & | \\ H & H & H & H \end{array} \qquad (26.1)$$

where $-R$ may be $-H$, $-CH_3$, $-Cl$, $-CN$, or a larger organic group. In polystyrene, for example, $-R$ is $-\hexagon$.

The polymer results from addition of the monomer units one to another to form the chain and there are no other products. There are many important addition polymers, of which polyethylene, polyvinyl chloride, and polystyrene are examples. Addition polymers are usually thermoplastic, which means they melt at high temperatures, under which conditions they can be molded or extruded into useful products.

In condensation polymers, linking of monomer units occurs along with elimination of a small molecule, usually water. The reaction is similar to that observed on formation of esters from acids and alcohols, which link up (condense) on elimination of water at the point of linkage. There are usually two different monomer units in condensation polymers, both of which have at least two reactive groups. Polyesters are condensation polymers obtained from two monomers, one of which is an acid with two or more −COOH groups and one an alcohol with two or more −OH groups. A general equation for polyester formation would be:

$$n\,[H-O-R-O-H \;+\; H-O-\underset{\underset{O}{\|}}{C}-R'-\underset{\underset{O}{\|}}{C}-O-H]$$

$$\rightarrow -R-O-\underset{\underset{O}{\|}}{C}-R'-\underset{\underset{O}{\|}}{C}-O-R-O-\underset{\underset{O}{\|}}{C}-R'-\underset{\underset{O}{\|}}{C}-O- \;+\; 2n\,H_2O \qquad (26.2)$$

As with addition polymers there are many possible R and R′ groups. In Dacron, for example, −R− and −R′− are −CH_2−CH_2− and − ⬡ − respectively. This material forms strong fibers and can also be pressed into a film, called Mylar. In this experiment we will make the polyester called glyptal, from glycerol, the trihydroxy alcohol obtained from fats, and phthalic acid, a dicarboxylic acid.

glycerol   ortho-phthalic acid

Another important group of condensation polymers is that of the polyamides, or Nylons. In these polymers one of the monomers is similar to the alcohol in Equation 26.2, except that instead of −OH groups it has −NH_2 groups. The other monomer is a dicarboxylic acid as in Equation 26.2. On condensation, water is eliminated and the polymer obtained, which is a polyamide, has the structure:

$$-R-\underset{\underset{H}{|}}{N}-\underset{\underset{O}{\|}}{C}-R'-\underset{\overset{\|}{O}}{C}-\underset{\overset{|}{H}}{N}-R-\underset{\underset{H}{|}}{N}-\underset{\underset{O}{\|}}{C}-R'-\underset{\overset{\|}{O}}{C}-\underset{\overset{|}{H}}{N}-$$

The similarity to polyesters is clear, but the properties of the two kinds of polymers differ, making both useful for particular purposes. The most famous polyamide is Nylon 6-6, in which $-R-$ is $-(CH_2)_6-$ and $-R'-$ is $-(CH_2)_4-$. This polymer was discovered by Wallace Carothers at the research laboratories of the du Pont Company in the late 1930's. The success of nylon, like that of mauve, resulted in a great expansion of industrial chemical research. In this experiment we will make Nylon 6-10 from hexamethylene diamine, $H_2N-(CH_2)_6-NH_2$, and sebacic acid, $HO-\underset{\substack{\| \\ O}}{C}-(CH_2)_8-\underset{\substack{\| \\ O}}{C}-OH$.

PRELIMINARY STUDY

1. Review Chapter 16 in the text.
2. (a) Write the structural formula of the monomer used in making polyethylene.
    (b) Write the structural formula of the polymer formed by the polymerization of styrene,

    (c) Write the structural formulas of the monomers used in making Dacron and Nylon 6-6.

PROCEDURE

Wear your safety goggles, apron, and gloves while performing this experiment. Avoid contact with any of the chemicals and their vapors.

Part I: Preparation of polymethylmethacrylate.

Lucite and Plexiglas are commercial names for polymethylmethacrylate. This addition polymer is noted for having a transparency similar to glass. It is thermoplastic and can be melted and formed into numerous shapes. The monomer, methyl methacrylate, contains an ester functional group and a carbon-carbon double bond.

methyl methacrylate

(a) Half-fill a 400 ml beaker with water and bring it to a boil. Place 5 ml of methyl methacrylate in a small test tube and add a small amount (about 0.05 g) of benzoyl peroxide. Stir the mixture until all of the benzoyl peroxide dissolves. Heat the tube and contents in the water bath for 10–15 minutes. (Note: While waiting you may begin Procedure 2.)

> CAUTION: Avoid contact with methyl methacrylate and benzoyl peroxide. If you get any of either on your skin or clothing, wash it off with soap and water immediately.

(b) After the reaction mixture becomes viscous, remove the tube from the water bath to cool. Examine the polymer when cool.

(c) Reheat the polymer in the water bath and note if it softens. (Note: See your instructor for instructions on disposal of the tube and polymer.)

Part II: Preparation of glyptal.

Glyptal is the commercial name for a polymer made by condensation reactions between hydroxyl groups of glycerol and carboxyl groups of phthalic acid. Glyptal is classified as a polyester. In this experiment, phthalic anhydride is used in place of phthalic acid because of its greater reactivity. The net result of the reaction is the same as if the acid was used.

(a) Place 1 ml of glycerol, 3 g of phthalic anhydride and 0.1 g of sodium acetate, $NaC_2H_3O_2$, into a small test tube. Stir the mixture until a solution is formed. Heat in the boiling water bath from Procedure 1 for about ten minutes. After the mixture has become quite viscous pour it onto a glass plate and let it cool.

(b) After the product has hardened, scrape it off the plate with a spatula. Put the product in a test tube and dissolve it in 5 ml of methyl ethyl ketone (butanone). Coat a wooden splint with the glyptal solution and note the results after it has dried.

Part III: Preparation of a Nylon.

Nylon is a polyamide normally made from a diamine and a dicarboxylic acid. In the synthesis employed in this experiment, sebacoyl chloride is used instead of sebacic acid because of its greater reactivity. In this case a molecule of HCl is eliminated in each condensation reaction in place of the usual molecule of H_2O. The polymer has the same structure as it would if we used sebacic acid. The structural formulas of the monomers are:

$$
\begin{array}{cc}
\underset{\text{hexamethylenediamine}}{
\begin{array}{c}
H \\
\backslash \\
N - (CH_2)_6 - N \\
/ \qquad\qquad\quad \backslash \\
H \qquad\qquad\qquad H
\end{array}}
&
\underset{\text{sebacoyl chloride}}{
\begin{array}{c}
O \qquad\qquad\qquad O \\
\backslash\backslash \qquad\qquad\quad // \\
C - (CH_2)_8 - C \\
/ \qquad\qquad\qquad \backslash \\
Cl \qquad\qquad\qquad Cl
\end{array}}
\end{array}
$$

(a) Pour 20 ml of a sebacoyl chloride solution (in CCl_3CH_3) into a 100 ml beaker. Carefully pour 10 ml of a hexamethylene-diamine solution on top of the first solution so that it "floats". A film of nylon polymer will form at the interface of the solutions.

(b) Grasp the film with a tweezers and pull away from the beaker. A nylon "rope" will form. Lay the rope on a paper towel and continue to pull new rope. (If a rotating drum is available the rope may be wound continuously around the drum.) Put the nylon rope in a beaker and wash it with water before touching it with your hands. Allow the rope to dry and examine the product. (Note: any unused mixture should be stirred until no further polymer forms.) Wash your hands on completion of the experiment.

CALCULATIONS AND QUESTIONS

1. (a) Draw the structure of a section of a polymethylmethacrylate molecule containing three monomer units (see Eqn. 26.1). To what class of polymer does it belong?

   (b) What are the properties of the polymer which give it commercial value? Cite three commercial uses of the polymer.

2. Draw the structure of a section of the polymer formed by glycerol and phthalic acid (see Eqn. 26.2). Use two molecules of each monomer and assume that only the end hydroxyl groups of glycerol react. To what class of polymer does it belong? Why?

3. Draw the structure of a section of the polymer formed from sebacoyl chloride and hexamethylenediamine. Use two molecules of each monomer. To what class of polymer does it belong?

4. Consider a polymethylmethacrylate molecule containing 100,000 monomer units:

   (a) Calculate the molecular mass of the polymer.

   (b) Calculate the extended length of the polymer molecule in meters if the carbon-carbon bond length in the chain is 0.15 nm.

EXTENSIONS

1. Properties of Polystyrene

   (a) Cut up a polystyrene foam coffee cup and test its solubility in several organic solvents such as ethyl alcohol, acetone, and toluene. In which solvent is the polymer most soluble? Why?

   (b) Using the best solution from above, pour about a ml of it onto the surface of water in a beaker. After the solvent evaporates pick up the film with a stirring rod and note its properties.

   (c) Heat pieces of polystyrene foam in a test tube over a low flame and note whether it melts or decomposes.

   (d) Using tongs, heat a piece of polystyrene foam in a flame and note its combustibility. What are the products?

2. It is possible to make Nylon from just one monomer instead of two. Suggest the monomer one would use to make Nylon 6. Draw the structural formula of a section of the molecule containing 3 monomer units.

3. In glyptal the ester chains are sometimes cross-linked by oxygen, $-O-$, bridges, formed when the extra $-OH$ groups of glycerol react, eliminating H_2O. The resin is not thermoplastic and is very resistant to solvents. Draw the structural formulas of portions of two chains in a glyptal polymer showing two cross-links.

RATES OF CHEMICAL REACTIONS

OBJECTIVES

1. To become familiar with the concept of reaction rate.
2. To study the factors which affect the reaction rate.
3. To determine the concentration of an unknown based on its reaction rate.

DISCUSSION

The rate of a chemical reaction is a measure of the amount of a product that is produced during a given length of time. Ordinarily, we evaluate the rate by finding the change in concentration of product and dividing by the time it takes for that change to occur. If, in 50 seconds, the change in concentration of product was 0.010 moles/liter, then the rate of the reaction would be (0.010 moles/liter)/50 seconds, or 2.0×10^{-4} moles/liter-second. If, in a series of experiments, we measured the time required for a given, constant amount of product to form, under various reaction conditions, we could take the relative rate of the reaction to be simply the reciprocal of the time interval we measured. The relative rate of a reaction that took 50 seconds would be 1/50 second, or 0.020 sec^{-1}. In this experiment we will measure relative rate by taking the reciprocal of the time required for a reaction to occur.

Rates, and relative rates, of reactions depend on several factors. These factors include the nature of the reactants, their concentrations, the temperature, and the presence of catalysts. Some reactions go fast, some are slow, some occur at rates that are easy to measure in a laboratory period. Nearly all reactions go faster if the reactant concentrations are increased or if the temperature of the reaction mixture is raised. Catalysts are substances which influence the rate of a reaction but are not used up in the reaction.

The reaction we will study in this experiment is a simple one, the decomposition of thiosulfate ion, $S_2O_3{}^{2-}$, in acidic solutions. The thiosulfate ion breaks down in the presence of the H^+ ion in acidic solutions according to the equation:

$$S_2O_3{}^{2-}(aq) + 2\ H^+(aq) \rightarrow S(s) + H_2SO_3(aq) \tag{27.1}$$

139

Solid sulfur is a product of the reaction. If a solution of $S_2O_3^{2-}$ is mixed with one containing H^+ ion at moderate concentrations, nothing seems to happen for a while. Finally, the solution turns cloudy, as a colloidal precipitate of sulfur forms. At that point a certain fixed amount of product has been formed, so the time required for the cloudiness to appear is a measure of the rate of the reaction. We will measure the reaction time required for different concentrations of $S_2O_3^{2-}$, holding the temperature and H^+ ion concentration constant. From the reaction times we will calculate relative rates for the reaction, and see how these depend on the concentration of $S_2O_3^{2-}$ ion. We will then measure the time required at different temperatures, holding both reactant concentrations fixed. This will allow us to observe the effect of temperature on reaction rate.

PRELIMINARY STUDY

1. Review Chapter 17 in the text.
2. Practice problems:
    (a) If 3.0 ml of 0.10 M $Na_2S_2O_3$ are diluted with water to a total volume of 10.0 ml, what is the molarity of $S_2O_3^{2-}$ ion in the final solution? (Ans. 0.030 M)
    (b) If the solution from (a) is mixed with 10.0 ml of 1 M HCl, what will be the new molarity of $S_2O_3^{2-}$ ion? (Ans. 0.015 M)

PROCEDURE

Students should work in pairs on this experiment.Wear your safety goggles, apron, and gloves while performing this experiment.

Part I: The effect of concentration on reaction rate.

1. (a) Fill three 50 ml burets with 1 M HCl, 0.10 M $Na_2S_2O_3$, and distilled water, one liquid to a buret. Drain a little liquid from each buret to fill the tip and then refill until the level of each liquid is at 0.0 ml. Your teacher will show you how to read the level in a buret.
    (b) Obtain two racks of regular size test tubes, each with five tubes. To each of the tubes in one of the racks add 10.0 ml 1 M HCl. Label the tubes in the other rack from 1 to 5, either with labels or by remembering the position of each tube in the rack. Fill those tubes as directed in Table 27-1, with the indicated volumes of 0.10 m $Na_2S_2O_3$ and distilled water. The final prepared solutions will have concentrations of $S_2O_3^{2-}$ ion that you can calculate from the way you made the solutions up. Record the room temperature.

TABLE 27–1

TUBE NO.	VOLUME 0.10 M $Na_2S_2O_3$ IN ML	VOLUME DISTILLED WATER IN ML
1	10.0	0
2	8.0	2.0
3	6.0	4.0
4	4.0	6.0
5	2.0	8.0

(c) In this step we will carry out the reaction between $S_2O_3^{2-}$ ion in each of the five solutions you prepared with the H^+ ion in the solutions of HCl. The concentration of H^+ ion will have the same value in each reaction, since we have the same volume of 1 M HCl in each of the test tubes. The concentration of $S_2O_3^{2-}$ will be different in each reaction, since we used different volumes of the stock solution in each tube.

When all the solutions have been prepared, take tube no. 1 and one of the tubes containing 1 M HCl and mix the two solutions. Begin timing. Pour the mixture from one tube to the other four times to ensure that the solution is well mixed. Watch the solution carefully, looking down through the tube, and when you see a definite cloudiness appear, record the elapsed time. Repeat the experiment with each of the solutions in tubes no. 2 to 5.

(d) Obtain a 10.0 ml sample of a solution of $Na_2S_2O_3$ of unknown molarity. Measure the time it takes for this solution to produce a cloudiness when mixed with 10.0 ml of 1 M HCl.

Part II: The effect of temperature on reaction rate.

The effect of temperature on the rate of the reaction will be determined by measuring the times required for reaction at 35°C and 10°C and comparing them with the time observed at room temperature for tube no. 1 in Part I.

1. Prepare a warm water bath (35°C) in a 400 ml beaker, by judicious mixing of hot and cold tap water. Prepare a similar cold water bath (10°C) using ice and cold tap water. Into each of two test tubes put 5.0 ml 1 M HCl; place one of the tubes in each of the water baths. Into two other test tubes put 5.0 ml of 0.10 M $Na_2S_2O_3$. Put one of these tubes in each water bath.
2. After the tubes have been in the baths for at least five minutes, take the two hot tubes and mix the solutions they contain, beginning to measure time at that point. Pour the mixture from one tube to another four times and put the tube containing the reaction mixture back in the water bath. Record the time elapsed when the cloudiness becomes definite. Record the temperature of the bath. Repeat the experiment, using the tubes in the cold water bath. Although the volumes of reagents used are smaller than in Part I, the mixtures we prepare have the same concentrations as in the reaction in tube no. 1, so the times obtained can be compared to that for tube no. 1 in Part I.

SAMPLE DATA TABLE   EXPERIMENT 27

PART I

Tube No.	Time (sec)	Relative rate 1/time	Concentration of $S_2O_3^{2-}$	
			As prepared	Mixed with HCl
1			0.10 M	0.050 M
2				
Room temperature		_____ °C	Unknown no. _____	

PART II

Temp	Time (sec)	1/time (1/sec)
10°C		
35°C		

CALCULATIONS AND QUESTIONS

1. (a) Calculate the relative rate of each reaction in PART I by taking the reciprocal of each time measurement. Record your results in the data table.

   (b) In each reaction the concentration of H^+ ion in the reaction mixture has the same value. In the reaction mixtures $[H^+]$ is 0.5 M, since the HCl solution is mixed with an equal volume of the other reagent. In Reaction 1, the concentration of $S_2O_3^{2-}$ ion is 0.10 M in the tube, and, after mixing with HCl solution, conc $S_2O_3^{2-}$ becomes 0.050 M. The other tubes contain less $Na_2S_2O_3$, so the concentrations of thiosulfate ion in those tubes are lower than 0.10 M; after mixing with HCl they are lower still. For tubes no. 2 to 5, calculate conc $S_2O_3^{2-}$ as prepared in the tube, and after mixing with HCl. Recall that, for dilution of stock solutions,

$$V_1 \times M_1 = V_2 \times M_2$$

   where 1 refers to the stock solution and 2 to the prepared solution. Record your results in the data table.

2. (a) Make a graph on which you plot the relative rate of each reaction on the y-axis (ordinate) against the concentration of $S_2O_3^{2-}$ in the reaction mixture on the x-axis (abscissa). Both scales should begin at 0. Draw a smooth curve through the points, minimizing the distances from the points to the curve.

   (b) How does the relative rate of the reaction vary with concentration of $S_2O_3^{2-}$ ion? Can you make a mathematical statement relating the rate to concentration of thiosulfate ion? If the concentration of $S_2O_3^{2-}$ ion is doubled, what happens to the relative rate of the reaction?

   (c) On the basis of the time you measured, what was the concentration of

$S_2O_3^{2-}$ in your unknown solution? (Note: The graph will give you conc $S_2O_3^{2-}$ after mixing with HCl.)

3. (a) Calculate the relative rates of the reactions in PART II.
    (b) Make a graph on which you plot relative rate at 10°C, room temperature, and 35°C on the y-axis against reaction temperature, from 0°C to 50°C, on the x-axis. Draw a smooth curve through the data points, extending the curve to 0°C and 50°C.
    (c) Comment on the dependence of reaction rate on temperature.
    (d) On the basis of your graph, how long would it take for the reaction studied in Part II to occur at 0°C? at 50°C?

EXTENSIONS

1. Zinc metal reacts with hydrochloric acid, HCl, evolving H_2 gas and producing Zn^{2+} ions in solution. Devise an experiment to find out how the rate of the reaction depends on H^+ ion concentration. Note that the rate will also depend on the surface area of the metal. Carry out the experiment, and plot the data obtained as you did for the reaction of $S_2O_3^{2-}$ ion. If the $[H^+]$ is doubled, what happens to the relative rate of the reaction?

SYSTEMS IN CHEMICAL EQUILIBRIUM

OBJECTIVES

1. To understand what is meant by chemical equilibrium.
2. To observe the effects of concentration and temperature changes on systems in chemical equilibrium.

DISCUSSION

Some chemical reactions proceed essentially to completion. A piece of magnesium, ignited in oxygen, burns until all of the magnesium has been converted to magnesium oxide. Other reactions proceed for a while, but appear to stop before any of the reactants are all consumed. If carbon monoxide is mixed with steam at high temperatures, a reaction will occur to form carbon dioxide and hydrogen, but no matter how long you wait, not all of either reactant will be used up:

$$CO(g) + H_2O(g) \rightleftarrows CO_2(g) + H_2(g) \qquad (28.1)$$

If we start with a mole of CO and a mole of H_2O we might end up with 3/4 mole CO_2 and 3/4 mole H_2, and 1/4 mole CO and 1/4 mole H_2O left unreacted. The reaction clearly does not proceed to completion, but to some state where the reactants and products can, and do, coexist stably. In such a state the system is said to be in chemical equilibrium. This sort of behavior is very common with chemical reactions.

If we have a system in chemical equilibrium, we can alter the position of the equilibrium, that is, the extent to which reactants are converted to products, by changing the conditions. In all cases the system will respond in such a way as to counteract the change we try to make. If we increase the concentration of a reactant, the reaction goes to the right, forming products and thereby decreasing to some extent the concentration we tried to increase. If we add a product to the equilibrium mixture, the reaction occurs to the left, so as to decrease the concentration of the product that was added. Similarly, if we try to raise the temperature of the system, reaction will occur in such a way as to absorb heat. That is, the reaction will proceed in the endothermic direction. Reaction 28.1 is exothermic as written; ΔH is equal to -9.9 kcal. If we heat the equilibrium system, the reaction will occur to the left, since in that direction the reaction is endothermic. The mixture we cited earlier might end up with 1/2 mole of each of the reactants and products in the equilibrium system at some higher temperature.

In this experiment we will set up several equilibrium systems, using colors of reactants and products to detect the equilibrium state. We will then alter concentration or temperature and observe the direction in which the reaction shifts as the system goes to its new equilibrium state.

PRELIMINARY STUDY

1. Review Sections 18.1 and 18.4 in the text.
2. Given the reaction: $A^+(aq) + B^-(aq) \rightleftharpoons C(aq)$ $\Delta H = -20$ kcal
    If C is a desired product, and B^- is an expensive reactant, how could you maximize the amount of B^- that is converted to product under equilibrium conditions?

PROCEDURE

Wear your safety goggles, apron, and gloves while performing this experiment.

PART I: Equilibrium in acid-base indicators

Some dyes undergo remarkable color changes when the concentration of H^+ ion is changed in the solution in which the dye is present. These color changes can be used to determine the $[H^+]$ in the solution, so the dyes are called acid-base indicators. We will use acid-base indicators for this purpose in the next experiment. In this experiment we will investigate the equilibrium behavior of bromthymol blue, a common acid-base indicator.

In acidic solution, at relatively high H^+ ion concentrations, bromthymol blue is yellow. In basic solution, where $[H^+]$ is low, bromthymol blue is blue. At a certain $[H^+]$, equal amounts of the blue and yellow forms are present, and the solution is green. Denoting the blue and yellow forms by B^- and Y respectively, we can write the reaction of bromthymol blue with H^+ ion in the following way:

$$H^+(aq) + B^-(aq) \rightleftharpoons Y(aq) \tag{28.2}$$

The system can be in equilibrium at any concentration of H^+ ion, but only over a relatively small range of $[H^+]$ will there be appreciable amounts of both the blue and yellow forms present.

1. Pour about 5 ml of distilled water into a regular size test tube. Add two or

three drops of bromthymol blue to the water and mix. Then add 5 drops of 0.1 M HCl and stir. This will increase the $[H^+]$. Observe the color of the solution. Now, drop by drop, add 0.1 M NaOH to the solution, mixing after each drop. This will decrease the $[H^+]$, since the OH^- ions in the NaOH solution will react with H^+ ions, forming water. Note any color changes that occur. Continue to add the NaOH solution until no further color change occurs.

2. To the solution from Procedure 1 add 0.1 M HCl, drop by drop, until the solution changes color again. Note that by changing the conditions in the equilibrium system, we can easily force the reaction to go in either direction, converting the indicator from one form to the other. Reactions like this are said to be reversible. All reactions that go to equilibrium are in principle reversible.

PART II: The $Cr_2O_7^{2-}$–CrO_4^{2-} equilibrium system

The $Cr_2O_7^{2-}$ ion is orange in solution and the CrO_4^{2-} ion is yellow. Addition of H^+ ion will tend to convert CrO_4^{2-} to $Cr_2O_7^{2-}$ according to the following reaction:

$$2\ H^+(aq) + 2\ CrO_4^{2-}(aq) \rightleftarrows Cr_2O_7^{2-}(aq) + H_2O \qquad (28.3)$$

1. Pour 5 ml 0.1 M $K_2Cr_2O_7$ into a regular test tube. Convert the $Cr_2O_7^{2-}$ ion to CrO_4^{2-} by selecting the proper reagent from the following pair and adding it to the dichromate solution; the two reagents are 1 M HCl and 1 M NaOH.
2. When you have obtained the chromate ion in solution convert it back to dichromate by selecting the proper reagent from the pair in 1.
3. Using the solution obtained in 2, prepare, by addition of the proper reagent, a solution in which the concentrations of $Cr_2O_7^{2-}$ and CrO_4^{2-} appear to be equal. You can most easily do this by comparing the color of the solution you are making to the color of a solution containing only $Cr_2O_7^{2-}$ ions and the color of a solution containing only CrO_4^{2-} ions. The color of the solution you prepare should be intermediate between those of the two reference solutions.

PART III: Equilibria in solutions containing Co^{2+} ion

In water solution the Co^{2+} ion is pink. The color is due to the $Co(H_2O)_6^{2+}$ ion, which is the form in which the cobalt ion ordinarily exists in water. If Cl^- ions are present in high concentrations, it is possible to convert the $Co(H_2O)_6^{2+}$ ion to a species containing Cl^- ions, $Co(H_2O)_4Cl_2$, which is blue:

$$Co(H_2O)_6^{2+}(aq) + 2\ Cl^-(aq) \rightleftarrows Co(H_2O)_4Cl_2(aq) + 2\ H_2O \qquad (28.4)$$

$$\text{pink} \qquad\qquad\qquad\qquad \text{blue}$$

The two colored Co^{2+} species can be converted one to the other by appropriate changes in the concentration of Cl^- ion or of water and by changes in temperature.

1. (a) Prepare a solution of $Co(H_2O)_4Cl_2$ by dissolving 0.5 g $CoCl_2 \cdot 6H_2O$ in 10 ml of 6 M HCl in a 100 ml beaker.
   (b) Prepare a solution of $Co(H_2O)_6{}^{2+}$ ion by dissolving 0.5 g $CoCl_2 \cdot 6H_2O$ in 25 ml of water in a 100 ml beaker.
2. (a) To the solution prepared in 1(a), add water slowly. When the color change appears to be complete, put the beaker on a wire gauze/iron ring, and bring the solution to the boiling point. Record your observations. Turn off the burner and let the beaker cool on the lab bench. Note any changes which occur.
   (b) Suggest two ways that the solution prepared in 1(b) might be made to turn blue. Test your proposals experimentally on 5 ml portions of the solution. Do not add more than 5 ml of any reagent, and heat any solutions in a beaker, not a test tube.

SAMPLE DATA TABLE	EXPERIMENT 28
Experimental procedure	Observations
PART I 1	

QUESTIONS

1. In PART I, Procedure 1, as NaOH is slowly added:
   (a) What happens to the color of the solution?
   (b) Does the solution go through a state where appreciable amounts of both B^- and Y are present? Explain your reasoning.
   (c) What happens to $[H^+]$? to $[Y]$? to $[B^-]$?
2. In PART I, Procedure 2, apply Le Chatelier's principle to explain your observations of the behavior of the solution on addition of 0.1 M HCl.
3. (a) In PART II, Procedure 1, which reagent converts $Cr_2O_7{}^{2-}$ to $CrO_4{}^{2-}$? Explain why that reagent is effective by using Le Chatelier's principle.
   (b) Is the conversion of $Cr_2O_7{}^{2-}$ to $CrO_4{}^{2-}$ reversible? Explain.
4. (a) In PART III, Procedure 2(a), use Le Chatelier's principle to explain what happened on addition of water.
   (b) Is Reaction 28.4 reversible? exothermic? Explain your reasoning.
   (c) Explain your observations in Procedure 2(b).

EXTENSIONS

1. Determine the relative solubilities of $BaCrO_4$ and $BaCr_2O_7$ by adding 0.1 M $Ba(NO_3)_2$ to 2 ml portions of 0.1 M K_2CrO_4 and $K_2Cr_2O_7$ in separate test tubes. Using your results of PART II, develop a procedure for dissolving the less soluble compound.

2. The Cu^{2+} ion reacts with NH_3 to form the dark blue $Cu(NH_3)_4{}^{2+}$ ion.
   (a) To a 2 ml portion of 0.1 M $Cu(NO_3)_2$ in a test tube add 6 M NH_3 until the $Cu(NH_3)_4{}^{2+}$ ion forms.
   (b) Write the equation for the reaction that occurs.
   (c) To the solution from (a) add 6 M HCl until no further changes occur. Explain your observations as best you can. (Note that NH_3 will react with H^+ to form $NH_4{}^+$.)

MEASUREMENT OF pH WITH ACID-BASE INDICATORS

OBJECTIVES

1. To prepare a set of pH indicator standards.
2. To determine the pH of an unknown solution.
3. To find the equilibrium constant, K_a, of a weak acid.

DISCUSSION

Acids are chemical compounds which in water solution produce significant concentrations of H^+ ion. Some acids with which we have worked in previous experiments include HCl, HNO_3, and H_2SO_4. In water solution these acids ionize essentially completely. The products are H^+ ions, which make the solution acidic, and negative ions:

$$HCl(aq) \rightarrow H^+(aq) + Cl^-(aq) \tag{29.1}$$

$$HNO_3(aq) \rightarrow H^+(aq) + NO_3^-(aq) \tag{29.2}$$

$$H_2SO_4(aq) \rightarrow H^+(aq) + HSO_4^-(aq) \tag{29.3}$$

Because their ionization reactions go to completion, these acids, and a few others, are called strong acids. In 1 M HCl, $[H^+]$ = 1 M, $[Cl^-]$ = 1 M, and the concentration of HCl molecules is about zero.

Most acids are not strong. This means that most acids do not ionize completely. Acids which do not ionize completely are called weak. As a general rule, weak acids in solution are ionized to the extent of only a few percent. For the weak acid HA in solution, the dissociation reaction

$$HA(aq) \rightleftharpoons H^+(aq) + A^-(aq) \tag{29.4}$$

will usually be in equilibrium when the concentrations of H^+ and A^- ions are of the order of a few percent of the concentration of HA molecules.

By the Law of Chemical Equilibrium, the concentrations of HA, H^+, and A^- in solution must satisfy the following condition:

$$\frac{[H^+] \times [A^-]}{[HA]} = K_a \tag{29.5}$$

where K_a is a constant, called the ionization constant of the acid. K_a has a particular numerical value characteristic of the acid HA. In any solution containing HA, the concentrations of HA, H^+, and A^- must have values which satisfy Equation 29.5. For a solution of HA of known molarity, M_{HA}, the value of K_a can be found if one can measure $[H^+]$ in the solution. In such a solution:

$$[H^+] = [A^-] \text{ and } [HA] \cong M_{HA} \text{ (only a little HA ionizes)} \tag{29.6}$$

From the measured value of $[H^+]$ we can find $[A^-]$ and from M_{HA} we obtain $[HA]$. Substitution into Equation 29.5 gives us the value of K_a.

Chemists express the degree of acidity in a solution either by giving the value of $[H^+]$ or by stating a related quantity called the pH. The relation between pH and $[H^+]$ is given by the following equation:

$$pH = - \log [H^+] \tag{29.7}$$

Since the logarithm of a number X is equal to the exponent to which 10 must be raised to give that number, we can say that $\log 10^2$ equals 2, and $\log 10^{-5}$ equals -5. If in a solution, $[H^+]$ is 1×10^{-5} M, the pH of the solution, by Equation 29.7, must be 5. If $[H^+]$ equals 0.1, or 10^{-1}, the pH will be 1, and so on. As you become more familiar with pH you will find it a convenient term to use in speaking of acidic solutions.

In the previous experiment we worked with an acid-base indicator, bromthymol blue, whose reaction with H^+ ion we can express by the equation:

$$Y(aq) \rightleftarrows H^+(aq) + B^-(aq) \tag{29.8}$$

In the equation, Y is the yellow form of the indicator, which is present in solutions where $[H^+]$ is relatively high, and B^- is the blue form, present when $[H^+]$ is low. If we add acid to a solution of B^-, we find that $[H^+]$ goes up, $[Y]$ goes up, and $[B^-]$ goes down, and if we add enough acid the solution turns yellow.

The reasons for this behavior become more apparent if we realize that Y is really just a weak acid, and that Equation 29.8 is just Equation 29.4 written in a slightly different formulation. This means that the weak acid Y has a dissociation constant K_a, and that a condition on $[Y]$ like that in Equation 29.5 must apply:

$$\frac{[H^+] \times [B^-]}{[Y]} = K_a \qquad \text{or} \qquad \frac{[B^-]}{[Y]} = \frac{K_a}{[H^+]} \tag{29.9}$$

If you look carefully at the right hand equation just above, it should be clear that the ratio of $[B^-]$ to $[Y]$ depends on the ratio of K_a to $[H^+]$:

If $K_a = [H^+]$, then $[B^-] = [Y]$ and the solution is green

   $K_a = 10 \times [H^+]$, then $[B^-] = 10 \times [Y]$ and the solution is essentially blue

   $K_a = 0.1 \times [H^+]$, then $[B^-] = 0.1 \times [Y]$ and the solution is essentially yellow

In our experiments with bromthymol blue, we forced the $[H^+]$ to be large or small as compared to K_a, and hence forced the reaction to produce the yellow or blue form of the indicator. Note that the change in color occurs over a rather small range of $[H^+]$ ($\sim$ 2 pH units). The indicator is most sensitive to $[H^+]$ when its value of K_a is actually equal to $[H^+]$. For bromthymol blue, K_a is equal to about 1×10^{-7}. In a solution containing bromthymol blue that is green, we can say that $[H^+]$ must be 1×10^{-7} M and the pH must be 7. Since all indicators work on the same principle as bromthymol blue, this approach allows us to use indicators to find $[H^+]$ and the pH of unknown solutions.

In this experiment we will prepare a set of solutions of known $[H^+]$, by making successive dilutions of a 0.1 M HCl solution. We will then add various acid-base indicators to these solutions, observing the change in color with changing $[H^+]$. Given these reference solutions, we will find the $[H^+]$ of an unknown solution of HCl and of a weak acid in its 1 M solution. From the last piece of data we will be able to calculate K_a for the weak acid.

PRELIMINARY STUDY

1. Review Chapter 19 in the text.
2. Practice Problem: Calculate the pH of a 0.001 M HNO_3 solution. (Ans: 3)
3. Review the procedure for writing the equilibrium constant expression, K_a, for a weak acid. Write the expression for K_a for hydrofluoric acid, HF.
4. If in a solution containing bromthymol blue, $K_a = 100\ [H^+]$, what color will the solution have?

PROCEDURE

Students may work in pairs on this experiment. Wear safety goggles, apron, and gloves.

1. (a) Obtain two test tube racks and 21 regular size test tubes. Add 35 ml 0.10 M HCl stock solution to a clean, dry 50 ml graduated cylinder. Pour 10 ml of the acid into each of three test tubes, and put them on a test tube rack where you can find them. Label the tubes or the position on the rack with the value of $[H^+]$, which will be 0.10 M since HCl is completely ionized in solution.

    (b) Check the level in the 50 ml graduate and make sure it is at 5.0 ml; add or take out a little 0.1 M HCl if necessary. Dilute with water until the volume is 50.0 ml. Pour the diluted solution into a beaker and stir for at least 10 seconds with a stirring rod. Pour about 10 ml of this solution into each of three test tubes, and put the test tubes on the rack, along with a label indicating the $[H^+]$ in the solutions.

(c) Rinse out the graduate with a few ml of the solution in the beaker, and then pour solution from the beaker into the graduate until the level is at the 5.0 ml mark. Dilute this solution as before, to 50.0 ml. Rinse out the beaker with distilled water and then pour the solution into the beaker, stirring well. Pour 10 ml of the solution in the beaker into three test tubes and label with $[H^+]$.

(d) Repeat step (c) two more times, to obtain two more dilutions of HCl. You should have, finally, five solutions of HCl, ranging in concentration from 1.0×10^{-1} M to 1.0×10^{-5} M, with 10 ml of each solution in three test tubes. Now, arrange the 15 test tubes into three groups of 5, with one test tube from each concentration in each group.

(e) To each solution in the first group add three drops of thymol blue. To each solution in the second group add three drops of methyl orange. Then add three drops of methyl red to each solution in the third set. Shake each tube well and record the color of each solution.

2. Obtain a 30 ml sample of hydrochloric acid of unknown concentration. Divide this solution equally among three test tubes. Add three drops of thymol blue to the first tube, three drops of methyl orange to the second, and three drops of methyl red to the third. Determine the hydrogen ion concentration in the unknown by comparing its colors with the three indicators to those of the groups of standard solutions prepared in 1. If you do not obtain a perfect match, estimate the $[H^+]$ in the unknown sample as best you can.

3. Obtain a 30 ml sample of a 1.0 M unknown weak acid. Determine the hydrogen ion concentration of the weak acid as in Procedure 2.

SAMPLE DATA TABLE   EXPERIMENT 29

Test Tube	$[H^+]$	pH	Indicator Color		
			Thymol Blue	Methyl Orange	Methyl Red
1	1.0×10^{-1} M	1			
2					

HCl Unknown

No. _____

Weak Acid

No. _____

CALCULATIONS AND QUESTIONS

1. (a) Calculate the pH of the standard solutions prepared in Procedure 1. Enter the values in the data table. What effect does a dilution by a factor of 10 have on the pH of an HCl solution?
   (b) Estimate the pH range over which each of the indicators changes color. State the color of the form present at low pH and at high pH. Estimate K_a for each indicator.
   (c) What was the pH of the unknown HCl solution? Give your reasons.
2. (a) What was $[H^+]$ in the sample of 1.0 M weak acid, whose formula you may take to be HA?
   (b) What was $[A^-]$? What was $[HA]$? (See Eqn. 29.6)
   (c) Substituting the values from (a) and (b) into the expression for K_a for the acid, calculate the value of K_a. (See Eqn. 29.5)
   (d) Assume that a salt of the weak acid Ha, such as NaA, is added to the weak acid solution. Using Le Chatelier's Principle, explain the effect that this would have on the $[H^+]$ and pH of the solution.

EXTENSIONS

1. In this experiment we made solutions in which $[H^+]$ was as low as 1×10^{-5} M by dilution. Why would it not be possible to make a solution in which $[H^+]$ was 1×10^{-8} M by dilution of the 1×10^{-5} M HCl solution you prepared?
2. Prepare a natural indicator from red cabbage. Cut up a few small leaves of red cabbage and boil in 50 ml of water until a purple solution is obtained. Using HCl and NaOH solutions, prepare 10 ml pH standards from pH = 1 to pH = 13. Add about 2 ml of cabbage juice to each solution. Record the pH and color of each solution. Can you find a value of K_a for red cabbage indicator?

PROPERTIES OF ACIDS AND BASES

OBJECTIVES

1. To determine the acidity or basicity of solutions of ionic substances.
2. To carry out some reactions of acids with bases.
3. To learn to write equations for acid-base reactions.

DISCUSSION

Most ionic substances in solution have acidic or basic properties in addition to the properties one would expect from their ions. A solution of $CuCl_2$, which contains Cu^{2+} and Cl^- ions, will also have acidic properties due to the interaction of Cu^{2+} ions with the water in which it is dissolved. Such interactions of ions with water are very common and must be recognized if one is to understand why solutions of ionic solids are usually acidic or basic.

An ion will behave as a base in solution if it can be associated with a weak acid. The following ions are basic because they can be converted, by addition of one or more H^+ ions, to the weak acid shown to their right:

BASIC ION	WEAK ACID	BASIC ION	WEAK ACID
F^-	HF	CO_3^{2-}	H_2CO_3
CH_3COO^-	CH_3COOH	PO_4^{3-}	H_3PO_4
CN^-	HCN	OCl^-	HOCl

An ion will behave as an acid in solution if it can be associated with a weak base. The following ions are acidic because they can be converted, by removal of a hydrogen ion, to the weak base shown to their right:

ACIDIC ION	WEAK BASE	ACIDIC ION	WEAK BASE
NH_4^+	NH_3	$Cu(H_2O)_4^{2+}$	$Cu(H_2O)_3OH^+$

An ion will be neutral in solution if it is derived from a strong acid or a strong base. The following ions are neutral because they arise from ionization of the strong acid or base shown to their right:

NEUTRAL ION	STRONG ACID	NEUTRAL ION	STRONG BASE
Cl^-	HCl	Na^+	NaOH
NO_3^-	HNO_3	K^+	KOH
Br^-	HBr	Ba^{2+}	$Ba(OH)_2$

A solution gets its acidity or basicity from the ions it contains. In solutions of ionic substances, the ions may be acidic, basic, or neutral. Usually at least one of the ions is neutral, so the other one fixes the pH of the solution. A solution of $CuCl_2$ contains the acidic Cu^{2+} ion (hydrated) and the neutral Cl^- ion, so the solution is acidic.

The reason ions like F^- are basic is that they interact with water, taking some H^+ ions from water and making a weak acid like HF. When this happens, some OH^- ions are left in the solution and it is basic:

$$F^-(aq) + H_2O \rightleftarrows HF(aq) + OH^-(aq) \qquad (30.1)$$

All basic anions interact in a similar way with water. Acidic ions are acidic because they tend to ionize a little, producing some H^+ ions and that makes the solution acidic. Neutral ions do not interact with water so they don't affect the pH of a solution.

When solutions that are acidic are mixed with solutions that are basic, a reaction occurs between the acidic and basic species. These may be H^+ ions or OH^- ions, or any of the ions we have listed which are acidic or basic. Below we have listed some pairs of solutions that might be mixed, the ions that are present in those solutions, and the reactions that occur on mixing:

SOLUTIONS MIXED	MAIN SPECIES PRESENT	REACTION THAT OCCURS
HCl and NaOH	(H^+) Cl^- Na^+ (OH^-)	$H^+(aq) + OH^-(aq) \rightarrow H_2O$
HCl and $NaCH_3COO$	(H^+) Cl^- Na^+ (CH_3COO^-)	$H^+(aq) + CH_3COO^-(aq) \rightarrow CH_3COOH(aq)$
CH_3COOH and NaOH	(CH_3COOH) Na^+ (OH^-)	$CH_3COOH(aq) + OH^-(aq) \rightarrow CH_3COO^-(aq) + H_2O$
HCl and Na_2CO_3	(H^+) Cl^- Na^+ (CO_3^{2-})	$2\,H^+(aq) + CO_3^{2-}(aq) \rightarrow H_2CO_3(aq)$
HCl and NH_3	(H^+) Cl^- (NH_3)	$H^+(aq) + NH_3(aq) \rightarrow NH_4^+(aq)$
NH_4Cl and NaOH	(NH_4^+) Cl^- Na^+ (OH^-)	$NH_4^+(aq) + OH^-(aq) \rightarrow NH_3(aq) + H_2O$

Note that although all of the reactions we have written would be classified as acid-base reactions, they do not all have the same equation. The reactions occur between the acidic and basic species present. These species are obtained by assuming that in solution all ionic substances and strong acids and bases are fully ionized, and that all weak acids and weak bases are essentially un-ionized. (NH_3 solutions are indeed basic, but the main species present is NH_3, not OH^-; CH_3COOH solutions are acidic, but the main species in solution is CH_3COOH, not H^+.)

One can show experimentally that reactions occur when acidic and basic solutions are mixed, by using acid-base indicators, which will change color when the pH changes. If one adds phenolphthalein to a solution of HCl and then slowly adds a solution of NaOH, when the reaction between H^+ and OH^- ions is complete the solution will turn red. Another way to detect a reaction is by the odor of reactants or products. Only molecular substances have an odor, ionic species do not. A solution containing NH_3 will smell of ammonia, and one containing acetic acid will smell of vinegar.

Insoluble solids containing basic ions will react with acidic solutions in the same way that the basic ions would react. Since the main species present is a solid, the reaction is written for the solid. The reaction that occurs when insoluble barium carbonate dissolves in a solution of HCl is written as:

$$BaCO_3(s) + 2\,H^+(aq) \rightarrow Ba^{2+}(aq) + H_2CO_3(aq)$$

If the concentration of H_2CO_3 gets too high, CO_2 gas will effervesce from the solution.

In this experiment we will carry out many of the reactions we have discussed here. If you use the principles we have presented, you will have no trouble understanding what is happening and will be able to write equations for the reactions you observe.

PRELIMINARY STUDY

1. Predict whether 0.1 M solutions of the following substances will be acidic, basic, or neutral: HNO_3, KCl, NaCN, NH_4Br
2. Write equations for the reactions which occur when solutions of the following are mixed: HNO_3 and NaOH; HCl and NaF; HBr and NH_3.

PROCEDURE

Wear your safety goggles, apron, and gloves while performing this experiment. In experiments where you need to shake a solution, stopper the tube with a cork, not your finger.

1. (a) Dissolve about 0.1 g of each of the following substances in separate regular test tubes containing about 5 ml of distilled water, one substance per tube: NaCl, Na_2CO_3, NaOH, NH_4Cl, and $FeCl_3$.
   (b) Test each solution with red and blue litmus paper (or pH paper), and note whether the solution is acidic, basic, or neutral.
2. Pour 10 ml of 1 M HCl into a 50 ml beaker and add two drops of phenolphthalein indicator. Slowly add 1 M NaOH solution to the acid, while stirring, until a color change takes place. Then slowly add 1 M HCl until another change occurs.
3. (a) Pour 2 ml of a 1 M CH_3COOH solution into a small test tube and carefully note its odor. Add slightly more than 2 ml of a 1 M NaOH solution to the acetic acid. Shake the tube well and note the odor of the mixture.
   (b) Pour 2 ml of a 1 M $NaCH_3COO$ solution into a test tube and note its

odor. Add slightly more than 2 ml of a 1 M HCl solution to the sodium acetate solution. Shake the tube well and carefully smell the mixture.

4. (a) Pour 2 ml of a 1 M NH_3 solution into a test tube and carefully smell it. Add slightly more than 2 ml of a 1 M HCl solution to the ammonia solution. Shake the tube well and note the odor of the mixture.

   (b) Pour 2 ml of a 1 M NH_4Cl solution into a test tube and note its odor. Add slightly more than 2 ml of a 1 M NaOH solution to the ammonium chloride solution. Shake the tube well and carefully smell the mixture.

5. (a) Add a small marble chip, $CaCO_3$, to each of two test tubes, one containing 3 ml of 6 M HCl and the other 3 ml of 6 M CH_3COOH. Compare the rates of the two reactions. What is the gas being evolved?

   (b) To 2 ml of 0.1 M $FeCl_3$ in a test tube, add 1 M NaOH until a precipitate of $Fe(OH)_3$ is present after stirring. Then add 1 M HCl, with stirring, until no further change occurs.

SAMPLE DATA TABLE   **EXPERIMENT 30**

Solution   Indicator Test
_____

Procedures 2–5:
Reactants   Observation

QUESTIONS

1. For each solute of Procedure 1 whose solution tested acidic or basic, write a balanced chemical equation which explains its behavior.

2. Explain why one of the solutes of Procedure 1 produced a neutral solution. Give the formula of another salt which would also produce a neutral solution.

3. For each reaction mixture in Procedures 2, 3, and 4, write the formulas of the substances whose solutions are mixed. Then write the formulas of the main species present in those solutions. Circle those that are not neutral. Then write balanced equations for each of the reactions that occurred.

4. (a) How do you account for the difference in the reaction rates of hydrochloric acid and acetic acid in Procedure 5(a)?

   (b) Write a balanced equation for the reaction of calcium carbonate with hydrochloric acid solution. Be sure to use as reactants the species that actually participate in the reaction.

   (c) Write a balanced equation for the reaction of $Fe(OH)_3$ with 1 M HCl in Procedure 5(b).

EXTENSIONS

1. Sulfuric acid ionizes in two steps, the first one producing HSO_4^-. In that ionization H_2SO_4 behaves as a strong acid. The HSO_4^- ion also can ionize, but behaves as a weak acid. Would a solution of $NaHSO_4$ be acidic, basic, or neutral? How about a solution of Na_2SO_4? Check your predictions by experiment, and explain your observations as best you can.

IONIC PRECIPITATION REACTIONS

OBJECTIVES

1. To carry out a series of precipitation reactions.
2. To deduce the ions which precipitate when solutions are mixed together.
3. To identify the ions in an unknown solution from its precipitation behavior.
4. To learn how to write equations for precipitation reactions.

DISCUSSION

Ionic compounds consist of positive and negative ions. When such compounds dissolve in water, they tend to ionize completely, so their solutions contain the ions that were present in the crystalline solid. The equation for the reaction by which an ionic compound dissolves is easy to write, if you know the ions present in the solid. In the case of potassium nitrate, we would have simply:

$$KNO_3(s) \rightarrow K^+(aq) + NO_3^-(aq) \qquad (31.1)$$

When two solutions of different ionic compounds are mixed, ions of opposite charge may interact to form a precipitate. The precipitate will be produced if the compound containing those ions is essentially insoluble in water. If, for example, we mix a solution of $AgNO_3$ with one of KBr, the ions present would be Ag^+, NO_3^-, K^+, and Br^-. Since a precipitate forms, it will have to be either AgBr or KNO_3. The other two possible compounds, $AgNO_3$ and KBr, were used to make the original solutions and so are soluble. If we know, or can find out, that KNO_3 is soluble, as it indeed is, then we can be sure that the precipitate is AgBr. The equation for the precipitation reaction would be

$$Ag^+(aq) + Br^-(aq) \rightarrow AgBr(s) \qquad (31.2)$$

The K^+ and NO_3^- ions are not in the equation since they are not changed as a result of the mixing of the two solutions. To write the balanced equation for a precipitation reaction you need only know the formulas of the reacting ions and take account of the fact that the total charge on the positive ions must equal the total charge on the negative ions. The solid precipitate must have no net charge.

In this experiment you will be given a set of six aqueous solutions of ionic solids. By mixing each solution with each of the others, in pairs, you will in some cases obtain precipitates and in other cases there will be no reaction. From your observations you should be able to determine which ion pairs form precipitates and which do not. The solutions you will work with have been selected so that in no case will two precipitates form when two solutions are mixed. Having dis-covered the insoluble ionic substances, you should be able to establish their chemical formulas and write balanced equations for the reactions by which they precipitate from solution. Given an unknown solution, taken from the set with which you worked, you will be able to determine its identity by observing its precipitation behavior when mixed with solutions in the set.

PRELIMINARY STUDY

1. Review Section 20.1 in the text.
2. What are the formulas of the ions present in each of the following solutions:

   (a) 0.1 M K_2CrO_4 (b) 0.2 M $FeCl_3$ (c) 0.5 M NaOH   (d) 0.1 M $Al(NO_3)_3$

3. A 0.1 M $BaCl_2$ solution is mixed with 0.1 M $CuSO_4$ and a precipitate forms. Mixing 0.1 M KCl with 0.1 M $Cu(NO_3)_2$ produces no precipitate. Name six ion pairs which can form soluble ionic solids and one that cannot. What is the formula of the precipitate that is formed in this experiment?

PROCEDURE

1. (a) Obtain a set of six solutions and a spot plate. Record the ions present in each solution on your data table. (If spot plates are not available, use a glass plate or a plastic sheet. Mark the plate off into squares with a grease pencil.)

   (b) Mix 2–3 drops of one solution with 2–3 drops of another solution in the indentation of the spot plate. (Use only 1 drop of each solution if a flat surface is used.) Be careful not to touch the solution droppers to the reaction mixture or contamination of the solutions may result. Use a stirring rod if you wish to stir the mixtures. Describe the results of any reaction in your data table. A precipitate produces a cloudy or opaque mixture. Some precipitates form immediately while others take a few minutes. If *no reaction* takes place, mark NR in the data table space. Mark an x in the data table where a solution combination represents a duplicate test. Continue mixing pairs of solutions until you have tested all possible combinations.

2. Obtain an unlabeled sample of one of your six solutions. Test the unknown solution with your six solutions. Record the results in your data table and identify the solution.

SAMPLE DATA TABLE   EXPERIMENT 31

Ions in Solution	1.	2.	3.	4.	5.	6.
1. _____	X					
2. _____	X	X				
3. _____	X	X	X			
4. _____	X	X	X	X		
5. _____	X	X	X	X	X	
6. _____	X	X	X	X	X	X

Unknown No.

_____

QUESTIONS

1. (a) In each case where a precipitate was formed on mixing two solutions, write the formulas of the two compounds which might have precipitated.
   (b) Record all the ion pairs in those cases where mixing two solutions did not produce a precipitate. None of these ion pairs, nor the pairs in the original reagents, could be in the compounds you listed in (a). Identify each precipitate in (a) and put a circle around its formula.
2. Write the equation for the reaction by which each of the precipitates you observed in this experiment was produced. (One equation for each different substance.) Write the name of each insoluble compound below its formula.
3. Identify the substance present in the unknown solution you used in Procedure 2. What is the evidence you used to decide what it was?

EXTENSIONS

1. Repeat the experiment with a second set of solutions.
2. Determine the solubility and the solubility product constant of calcium sulfate. Add 2.00 g of $CaSO_4$ to one liter of water in an Erlenmeyer flask. Shake the mixture well and allow it to sit overnight to reach saturation equilibrium. Using weighed filter paper, filter out the excess solute. Dry the paper overnight in a hood or low temperature oven. Weigh the filter paper when it is dry, and calculate the amount of undissolved solute. The solid residue will be $CaSO_4 \cdot 2H_2O$. Using this information, calculate the solubility of $CaSO_4$ in moles/liter. From the solubility, find the solubility product, K_{sp}, for calcium sulfate.

ACID-BASE TITRATIONS

OBJECTIVES

1. To become familiar with the methods of volumetric analysis.
2. To standardize a basic solution and an acidic solution using acid-base titrations.

DISCUSSION

One of the important general methods used to determine the composition of unknown samples is volumetric analysis. As the name implies, volumetric analysis depends in part on accurate measurements of reagent volumes. These volumes are measured with volumetric apparatus, in particular, burets, pipets, and volumetric flasks. The actual experiment usually involves a titration of one reagent with another. In a titration one reagent is added from a buret to a known amount of another reagent. A chemical indicator is used to signal the point when the amounts of the two reagents are chemically equivalent. That is, when, in the reaction mixture, just enough reagent from the buret has been added to react exactly with the amount of the other reagent that is present in the system. From the data obtained in the titration one can determine the concentration or the number of moles of one of the reagents used from the known amount of the other reagent.

In this experiment we are concerned with reactions between acids and bases, which are sometimes called neutralization reactions. One of the reactions we will carry out is that which occurs when a solution of HCl is mixed with and neutralized by a solution of NaOH. You may recall from Experiment 30 that the reaction is between the H^+ ions from the HCl and the OH^- ions from the NaOH:

$$H^+(aq) + OH^-(aq) \rightarrow H_2O \qquad (32.1)$$

In the procedure we will put a known volume of a solution of HCl, whose molarity we wish to determine, into a flask. After adding a phenolphthalein indicator we will add a standardized solution of NaOH from a buret until the indicator turns pink. That will occur at the equivalence point, where, by Eqn 32.1,

$$\text{no. moles OH}^- \text{ added} = \text{no. moles H}^+ \text{ added} \qquad (32.2)$$

Relating number of moles to volume and molarity, we have

$$M_{OH^-} \times V_{NaOH} = M_{H^+} \times V_{HCl} \qquad (32.3)$$

Since the molarity of the OH^- and the NaOH are equal, as are the molarity of the H^+ and that of the HCl, we can write Eqn 32.3 in terms of the reagents:

$$M_{NaOH} \times V_{NaOH} = M_{HCl} \times V_{HCl} \qquad (32.4)$$

In the titration experiment we measure the volumes of the NaOH and the HCl and know the molarity of the NaOH. Hence we can use Equation 32.4 to find the molarity of the HCl and so standardize the HCl solution.

In the first part of the experiment we will standardize our NaOH solution by titration against a primary standard. A primary standard is a chemical substance of such purity that it can be used as a reference for standardizing other reagents. In acid-base titrations the usual primary acid standard is potassium hydrogen phthalate, an organic acid whose formula is $KHC_8H_4O_4$, or KHP for short. In water KHP is soluble and ionizes to form the K^+ and HP^- ions. The latter will react as a weak acid with the OH^- ions from an NaOH solution:

$$HP^-(aq) + OH^-(aq) \rightarrow H_2O + P^{2-}(aq) \qquad (32.5)$$

The mole ratio $HP^-:OH^-$ is clearly 1:1. We titrate a known mass of KHP with NaOH solution to a phenolphthalein end point. By Equation 32.5, at the end point,

$$\text{no. moles HP}^- = \text{no. moles OH}^- \qquad (32.6)$$

The number of moles of HP^- must equal the number of moles of KHP used. From the mass of KHP and its molar mass (204) we can find the number of moles of KHP in the sample. The number of moles of OH^- equals the number of moles of NaOH, so Equation 32.6 can be written in the form:

$$\frac{\text{Mass of KHP in sample}}{204 \text{ g/mole}} = M_{NaOH} \times V_{NaOH} \qquad (32.7)$$

From the mass of KHP and the volume of NaOH added to get to the end point, we can find the molarity of the NaOH solution by substitution into Equation 32.7.

This experiment, when done properly, is probably the most accurate of all those you have performed in this course. To get good results you must be careful with all measurements, and use proper technique. As you read over the Procedure, try to recognize those steps which must be done exactly right, since that will help you when you actually do the experiment.

PRELIMINARY STUDY

1. Practice Problems:
    (a) If 20.0 ml of 1.50 M NaOH are used in a titration how many moles of NaOH are used? (Ans: 3.00×10^{-2})
    (b) If the NaOH sample above is neutralized by 16.5 ml of HCl(aq) what is the concentration of the acid? (Ans: 1.82 M)
2. (a) Write the chemical equation for the neutralization reaction of NH_3(aq), a weak base, with HCl(aq).
    (b) Will the reaction mixture be acidic, basic, or neutral at the equivalence point? Hint: at the equivalence point the solution will contain NH_4^+ and Cl^- ions, so it will really be just a solution of NH_4Cl.

PROCEDURE

Wear your safety goggles, apron, and gloves while performing this experiment.

PART I: Standardization of an NaOH solution

1. To a clean, dry beaker add about 100 ml of the NaOH solution you will standardize. Thoroughly clean a buret by rinsing it with water several times. A clean buret will drain evenly, without leaving water spots on the glass wall. When the buret is clean, rinse the inside surface twice with a few ml of the NaOH solution to be standardized. This will insure that the NaOH concentration will be the same throughout the buret. With the stopcock closed, carefully fill the buret with the NaOH solution. Drain some of the solution until the tip of the buret is filled. Add NaOH solution to the buret until the liquid level is about at the 0 ml mark. Read and record the buret level to the nearest 0.1 ml.
2. Weigh out about 1 g of potassium hydrogen phthalate, $KHC_8H_4O_4$, to the nearest 0.01 g on a weighed piece of filter paper. Pour the solid into a clean 250 ml Erlenmeyer flask. Add about 30 ml of distilled water and swirl the flask until all of the solid dissolves. Wash any crystals which cling to the wall of the flask down into the solution with distilled water from a wash bottle. Add two drops of phenolphthalein indicator to the acid solution.
3. Place the flask under the buret, then lower the buret until its tip extends into the flask. Place a piece of white paper beneath the flask to help you see color changes. Titrate the solution, gradually adding base from the buret, swirling the flask to mix the solutions. Add base more slowly when the indicator color begins to persist. Stop titrating when the light pink color of the phenolphthalein remains for at least 30 seconds; you have reached the "end point." The change at the end point should be abrupt. The addition of a single drop of base should turn the solution from colorless to pink. Record the level of the NaOH solution remaining in the buret.
4. Refill the buret with NaOH solution to the 0 ml mark. Record the level in the buret. Repeat the titration (Procedures 2 and 3) with a second 1 g sample of $KHC_8H_4O_4$. Then refill the buret for PART II.

PART II: Standardization of HCl solution

1. To a clean, dry beaker add about 50 ml of the hydrochloric acid solution you
    will standardize. Thoroughly clean a second buret in the same manner as
    before. Rinse it twice with a little of the HCl solution and then fill it with the
    rest of that solution. Fill the tip by momentarily opening the stopcock. Read
    the level in the buret. Then deliver about 10 ml of the HCl solution into a
    clean 250 ml Erlenmeyer flask. Add 30 ml of water and 2 drops of phenol-
    phthalein indicator to the acid solution. Read the level in the NaOH buret.
    Titrate the acid with the NaOH solution to the pink end point. If you go past
    the end point, add a few drops of acid from the HCl buret and then carefully
    add NaOH until one drop causes the color to change to pink. Read the levels
    in the NaOH and in the HCl burets.
2. Repeat the titration with a second 10 ml sample of HCl. The ratios of volume
    NaOH to volume HCl used in the two titrations should agree to within 1
    or 2%.

SAMPLE DATA TABLE   **EXPERIMENT 32**

PART I: Run no. 1

mass of paper   _____g   vol. of NaOH (initial) _____ml

mass of paper + KHC$_8$H$_4$O$_4$ _____g   vol. of NaOH (final)   _____ml

mass of KHC$_8$H$_4$O$_4$ _____g   vol. of NaOH used   _____ml

PART II: Run no. 1

vol. of HCl (initial)   _____ml   vol. of NaOH (initial) _____ml

vol. of HCl (final)   _____ml   vol. of NaOH (final)   _____ml

vol. of HCl used   _____ml   vol. of NaOH used   _____ml

CALCULATIONS AND QUESTIONS

1. (a) Calculate the number of moles of KHC$_8$H$_4$O$_4$ (MM = 204) used in the
         standardization of the base.
    (b) Calculate the molarity of the NaOH solution (Eqn. 32.7).
    (c) Repeat the molarity calculation for the second run and calculate the
         average molarity of the base.
2. (a) Calculate the moles of NaOH used in the standardization of the acid.
         Use the average molarity of the NaOH as determined in Question 1(c).
    (b) Calculate the molarity of the HCl solution (Eqn. 32.4).

(c) Repeat the molarity calculation for the second run and calculate the average molarity of the acid.
3. In words, state what the calculation you made in Question 2(b) tells you about the HCl solution.

EXTENSIONS

1. (a) If one titrates a solution of NH_3 with a solution of HCl, the equivalence point will not be at pH 7. Why? Why would methyl orange be a better equivalence point indicator than phenolphthalein for this titration? (See results of Experiment 29.)
   (b) Standardize an NH_3 solution with your standardized solution of HCl. Use methyl orange as the equivalence point indicator. Make sure to check the colors of the indicator in acidic and basic solutions before beginning the titration.

TESTING CONSUMER PRODUCTS

OBJECTIVES

1. To determine the mass percent of acetic acid in vinegar.
2. To determine the mass percent of ammonia in household ammonia.
3. To determine the neutralizing capability of commercial antacid tablets.

DISCUSSION

In your home you use many commercial products which are acidic or basic in nature. Vinegar, fruit juices, carbonated beverages, baking soda, lye, household ammonia, and antacids are just a few examples. All of these products are regularly tested by their manufacturers as part of their quality control operations. Government agencies test such products for purity, uniformity of content, and agreement with label information and advertising claims.

In the previous experiment we titrated solutions of acids with bases in order to standardize solutions of HCl and NaOH. Those solutions can be used in this experiment to determine the acid or base content of some common household products.

Vinegar

Vinegar is a solution of acetic acid in water. Acetic acid is the most common organic acid, and like all organic acids, it is a weak acid. Vinegar can be neutralized by a solution of NaOH. The reaction is:

$$CH_3COOH(aq) + OH^-(aq) \rightarrow CH_3COO^-(aq) + H_2O \qquad (33.1)$$

If a measured volume of vinegar is titrated with a standardized solution of NaOH to a phenolphthalein end point, we can say, by Equation 33.1, that in the titration:

$$\text{no. moles } CH_3COOH = \text{no. moles } OH^- = \text{volume of NaOH} \times \text{molarity NaOH} \quad (33.2)$$
(in vinegar)

Hence, by the titration we can find the number of moles of acetic acid in the vinegar sample. Knowing the molecular mass of acetic acid, we can calculate the number of grams of acetic acid in the sample. Given the density of acetic acid, it is an easy matter to determine the mass per cent acetic acid in the vinegar:

$$\text{mass \% acetic acid} = \frac{\text{mass of acetic acid}}{\text{mass of vinegar sample}} \times 100 \quad (33.3)$$

Household Ammonia

Household ammonia is mainly a solution of ammonia, NH_3, in water. The solution may also contain some detergent and an organic solvent. Ammonia is a typical weak base and so can be neutralized with hydrochloric acid. The reaction that occurs is:

$$NH_3(aq) + H^+(aq) \rightarrow NH_4^+(aq) \quad (33.4)$$

If a measured volume of household ammonia is titrated with a standardized HCl solution to a methyl orange end point, we can say, by Equation 33.4, that in the titration:

$$\text{no. moles } NH_3 = \text{no. moles } H^+ = \text{volume HCl} \times \text{molarity of HCl} \quad (33.5)$$

If we wish to find the mass percent NH_3 in household ammonia, we can proceed as with the vinegar calculation, finding the mass of NH_3 from the number of moles NH_3 and its molecular mass. The mass of the sample can be determined from its volume and density.

Antacids

Commercial antacids contain one or more weak bases as their active ingredients. The most common weak base in these products is sodium hydrogen carbonate, $NaHCO_3$, although several other substances are also used. The weak bases are designed to neutralize "stomach acid." Stomach acid is secreted as approximately 0.1 M hydrochloric acid. It is generally believed that "acid indigestion" results when the pH of the stomach fluids falls below 3.

The weak bases in an antacid tablet will react with hydrochloric acid. If we take the weak base to be X^-, the neutralization reaction would be:

$$X^-(aq) + H^+(aq) \rightarrow HX(aq) \quad (33.6)$$

In this case, direct titration of the antacid tablet with hydrochloric acid is not feasible, since the antacid is relatively insoluble and the end point of the titration is not readily observed. The procedure that is actually used is called a back-titration. It involves addition of a known, excess, amount of HCl solution to a known mass of tablet, and then titration of the excess acid with a standardized NaOH solution. In the back-titration, using standardized HCl and standardized NaOH:

$$\text{total moles H}^+ \text{ added} = \text{moles base in antacid} + \text{moles NaOH required} \qquad (33.7)$$

or

moles base in antacid
= (total volume HCl $\times$ molarity HCl) – (volume NaOH $\times$ molarity NaOH)

$$(33.7')$$

Using Equation 33.7' and the mass of antacid, we can determine the capacity of the antacid to neutralize acid. It can be expressed as moles of acid per gram of antacid, or per tablet of antacid, or per penny of cost. It is also sometimes stated in advertisements as grams of stomach acid per gram of antacid.

PRELIMINARY STUDY

1. Review the techniques and calculation methods for acid-base titrations used in the previous experiment.
2. Practice Problem: Determine the mass percent of acetic acid in a 1.0 M CH_3COOH solution. You may assume a solution density of 1.0 g/ml. (Ans: 6.0%)

PROCEDURE

Your instructor will tell you which parts of this experiment to perform. Wear your safety goggles, apron, and gloves while performing this experiment.

PART I: Analysis of vinegar

1. Using a pipet and suction bulb, measure out exactly 5 ml of vinegar into a clean 250 ml Erlenmeyer flask. Add about 30 ml of distilled water and two or three drops of phenolphthalein solution.
2. Fill a clean buret with the standardized NaOH solution from the previous experiment (rinse with the NaOH before filling). Following the procedure in that experiment, titrate the vinegar solution with the standardized base. As you approach the end point, add the NaOH drop by drop. Stop the titration when the solution takes on a pink color that persists for at least 30 seconds. With care, the color change will occur on addition of one drop of base. Record the volume of base used.
3. Repeat the titration with a second 5 ml sample of vinegar.

PART II: Analysis of household ammonia

1. Using a pipet and suction bulb, measure out exactly 5 ml of the ammonia solution into a clean 250 ml Erlenmeyer flask. Add about 30 ml distilled water and two or three drops of methyl orange indicator. The solution will turn yellow.
2. Fill a clean buret with the standardized HCl solution from the previous experiment (rinse with the HCl before filling). Following the procedure in that experiment, titrate the ammonia solution with the standardized acid. As you approach the end point, which is orange, add the HCl drop by drop. Stop the titration when the solution takes on an orange color. If you go too far the solution will become red. With care, the color change from yellow to orange will occur on addition of one drop of acid. Record the volume of acid used.
3. Repeat the titration with a second 5 ml sample of the household ammonia.

PART III: Analysis of an antacid tablet

1. Weigh accurately a commercial antacid tablet (Rolaids, Tums, and so on). Grind the tablet in a mortar, and divide the powder into two equal portions. Weigh one of the portions on a piece of filter paper which you have previously weighed. Save the other half of the tablet for later use.
2. Add the weighed antacid powder to a 250 ml Erlenmeyer flask. Fill a clean, rinsed buret with the standardized HCl solution from the previous experiment. Record the initial level of the solution, and then add about 50 ml of the acid to the powder in the flask. Record the final level in the buret. Swirl the flask to dissolve the solid. All of it may not dissolve, since there may be some insoluble components.
3. Add 4–5 drops of methyl orange indicator to the solution in the flask. The solution should turn red, since the HCl added should be in excess. If the solution is yellow, refill the HCl buret and add about 10 ml more acid, or more if necessary, measuring the volume carefully. Fill a clean, rinsed buret with the standardized NaOH solution from the previous experiment. Back-titrate the acidic solution in the flask with the base, using the procedure in the previous experiment. As you approach the end point, add the base drop by drop. The end point of the titration occurs when the solution turns from red to orange. One drop of base will cause the color change if the titration is done carefully. Record the volume of base required.
4. Weigh the other half of the powder obtained from the antacid tablet and repeat the back titration procedure.

SAMPLE DATA TABLE   EXPERIMENT 33

Analysis of Vinegar		*Analysis of Household Ammonia*	
vol. of vinegar	———— ml	vol. of NH_3 soln.	———— ml
M of NaOH	————	M of HCl	————
vol. of NaOH (initial)	———— ml	vol. of HCl (initial)	———— ml
vol. of NaOH (final)	———— ml	vol. of HCl (final)	———— ml
vol. of NaOH used	———— ml	vol. of HCl used	———— ml

Analysis of Antacid			
brand of antacid	————	M of HCl	————
mass of tablet	———— g	total vol. of HCl used	———— ml
mass of paper	———— g	vol. of NaOH (initial)	———— ml
mass of paper + antacid	———— g	vol. of NaOH (final)	———— ml
mass of antacid used	———— g	vol. of NaOH used	———— ml

CALCULATIONS AND QUESTIONS

1. (a) If you analyzed vinegar, calculate the number of moles of NaOH used to neutralize the sample (Eqn. 33.2).
   (b) How many moles of acetic acid were there in the vinegar sample?
   (c) How much does a mole of acetic acid weigh?
   (d) How many grams of acetic acid were there in the sample?
   (e) How much did the sample weigh? The density of vinegar is about 1.00 g/ml.
   (f) What is the mass % acetic acid in the vinegar (Eqn. 33.3)?
   (g) If you analyzed a second sample of vinegar, repeat the calculations in (a)–(f). Report the average mass % acetic acid.

2. (a) If you analyzed household ammonia, how many moles of HCl were required to neutralize the sample (Eqn. 33.5)?
   (b) How many moles of NH_3 were there in the sample?
   (c) What is the molar mass of NH_3?
   (d) How many grams of NH_3 were there in the sample?
   (e) How much did the sample weigh? The density of household ammonia is about 1.00 g/ml.
   (f) What is the mass % NH_3 in the household ammonia?

(g) If you analyzed a second sample of household ammonia, repeat the calculations in (a)–(f). Report the average mass % NH_3 in the sample.

3. (a) If you analyzed an antacid tablet, how many moles of HCl were added to the powder sample?

   (b) How many moles of NaOH were added during the titration?

   (c) How many moles of base were there in the antacid sample used (Eqn. 33.7')?

   (d) How many moles of HCl did the powdered antacid sample neutralize?

   (e) How many moles of HCl would be neutralized by one gram of antacid?

   (f) Repeat calculations (a)–(e) for the second sample. Calculate the average value for (e).

EXTENSIONS

1. (a) Does a Rolaids consume "47 times its own weight of excess stomach acid"? Assume that stomach acid is 0.1 M HCl and that its density is 1.00 g/ml. Use class data, if necessary, for this part and those that follow.

   (b) Does a Tums tablet "neutralize 1/3 more acid than a Rolaids tablet"? If this claim is true, how might it be countered by the Rolaids manufacturer?

   (c) Compare the cost in cents of neutralizing a mole of HCl with Rolaids and with Tums. (Hint: Find the number of tablets per mole HCl, and the cost per tablet for the two antacids.

DEVELOPING A QUALITATIVE ANALYSIS SCHEME

OBJECTIVES

1. To investigate the solubility properties of the Group 2 ions.
2. To learn to make flame tests.
3. To develop a scheme for qualitative analysis based on precipitation reactions.
4. To identify the ions in a Group 2 unknown.

DISCUSSION

In inorganic qualitative analysis a typical problem is to determine which ions are present in an unknown solution. The traditional method of analysis for metallic cations involves separation of the ions into groups on the basis of solubilities in different solutions and the subsequent identification of the ions in a group on the basis of their different chemical properties.

If a solution contains only one positive ion out of a limited set, one can often identify the ion through its solubility properties in different reagents. Since ions usually have characteristic solubility properties, one need only find a known ion which has solubility properties which match those of the unknown.

If more than one positive ion may be present in a sample the problem of identification becomes somewhat more difficult. To deal with such problems one ordinarily needs to work with acid-base reactions, complex ion formation reactions, and oxidation-reduction reactions, as well as precipitation reactions. In the next experiment we will use a standard scheme based on such reactions for the qualitative analysis of an unknown.

In this experiment you will develop a scheme for the qualitative analysis of the following four Group 2 ions: magnesium, calcium, strontium, and barium. The relative solubilities of these ions will be determined in the presence of sulfate, hydroxide, oxalate, and chromate ions. When solutions of the Group 2 ions are mixed with solutions containing the above negative ions, a precipitation reaction will occur if the ion pair can form a compound with low solubility:

$$M^{2+}(aq) + X^{2-}(aq) \rightarrow MX(s) \qquad (34.1)$$

or

$$M^{2+}(aq) + 2\,Y^-(aq) \rightarrow MY_2(s) \qquad (34.2)$$

$$M^{2+} = Mg^{2+}, Ca^{2+}, Sr^{2+}, Ba^{2+}; \qquad X^{2-} = SO_4{}^{2-}, C_2O_4{}^{2-}, CrO_4{}^{2-}; \qquad Y^- = OH^-$$

By examining the solubility properties of the ion pairs in these tests, you will be able to develop a simple scheme for analyzing an unknown solution that contains a single positive ion. If more than one positive ion may be present in your unknown, it will probably be helpful to use acid-base as well as precipitation reactions to accomplish the analysis.

Occasionally ions can be identified by flame tests as well as by tests based on chemical properties. A flame test consists of putting a clean wire, wet with the unknown solution, in a bunsen flame and observing the color of the flame that appears. Sodium has a particularly characteristic strong yellow flame test. Potassium has a much weaker violet-red flame, lithium a red flame, and copper has a green flame test. Among the Group 2 ions, barium, strontium, and calcium give moderately good flame tests. Care must be used in identifying ions through their flame tests, since two ions may have flames with similar colors.

PRELIMINARY STUDY

1. Review Sections 22.1 and 22.2 in the text.
2. Construct a solubility table, in advance of class, similar to that shown in the Sample Data Table.

PROCEDURE

Wear your safety goggles, apron, and gloves while performing this experiment.

1. Add about 1 ml of 0.1 M solutions of the nitrate salts of magnesium, calcium, strontium, and barium to four small test tubes, one solution to a tube. To each tube add about 1 ml of 1 M H_2SO_4. Shake or stir each tube so that the solutions are mixed. Record your results in your solubility table. Note whether a precipitate (ppt) forms, as well as any characteristics that might distinguish it.
2. (a) Repeat the above procedure testing a new set of solutions, adding 1 ml of 1 M NaOH to each solution.
    (b) Repeat using 1 ml of 0.25 M $(NH_4)_2C_2O_4$, ammonium oxalate.
    (c) Repeat using 1 ml of 1 M K_2CrO_4 plus 1 ml of 1 M acetic acid, CH_3COOH.
3. (a) Obtain a flame tester made of a piece of Nichrome or platinum wire in a glass holder. Clean the wire loop at the end of the tester by alter-

nately dipping it in 6 M HNO_3 and heating it in a bunsen flame. When clean, the only color emitted is the glow of the wire. Test for background flame color with a sample of the distilled water from which your solutions were made. Many water samples contain a small, detectable amount of sodium ion.

(b) Perform flame tests on the magnesium, calcium, strontium, and barium ion solutions. To do these tests, put about 2 ml of the solution containing the ion in a 50 ml beaker and carefully evaporate to near dryness. Dip the wire in the concentrated solution or the crystals and put it in the bunsen flame. Note the color and persistence of each flame. Clean the wire loop with 6 M HNO_3 between tests.

4. Devise a flow chart for the qualitative analysis of the magnesium, calcium, strontium, and barium ions. The scheme should separate the ions from solution one by one so that you can identify them if they are present. Note: If your procedure is to work for several ions, you will need to consider the pH of your solutions. In particular, hydroxides and oxalates are soluble in strongly acidic solution.

5. (a) Obtain a single ion unknown and determine its identity by following the procedure outlined in your flow chart. Confirm your results by performing a flame test on the original solution.

(b) Obtain a multiple ion unknown which may contain any number of the Group 2 ions. Determine the presence or absence of all ions by following the procedure in your scheme.

SAMPLE DATA TABLE   **EXPERIMENT 34**

1. Solubilities of Salts of the Alkaline Earths:

	1 M H_2SO_4	1 M NaOH	0.25 M $(NH_4)_2C_2O_4$	1 M K_2CrO_4 1 M CH_3COOH
$Mg(NO_3)_2$				
$Ca(NO_3)_2$				
$Sr(NO_3)_2$				
$Ba(NO_3)_2$				

2. Flame Tests:

Distilled H_2O	$Mg(NO_3)_2$	$Ca(NO_3)_2$	$Sr(NO_3)_2$	$Ba(NO_3)_2$

3. Single Ion Unknown No. _____   Ion present: _____

4. Multiple Ion Unknown No. _____   Ions present: _____

QUESTIONS

1. Write a balanced chemical equation for each precipitation reaction which separates ions in your flow chart.

2. (a) A solution containing the Li^+ ion produces a red flame color when heated. With what Group 2 ion would this flame test possibly conflict?

   (b) A solution containing Cu^{2+} ion produces a green flame color when heated. With what Group 2 ion would this flame test possibly conflict?

3. Name a single reagent which will separate the following pairs of ions:

   (a) Mg^{2+}, Ca^{2+} (c) Ca^{2+}, Sr^{2+}

   (b) Mg^{2+}, Ba^{2+} (d) Ca^{2+}, Ba^{2+}

EXTENSIONS

1. If you limit yourself to only precipitation reactions, it is likely that your scheme for the qualitative analysis of the Group 2 cations will fail, at least with some of the unknowns. The difficulty is that the reagents used are acidic or basic, and this property, as well as the ions present, must be taken into account. In particular, Ca^{2+} will not precipitate with $C_2O_4^{2-}$ ion if the solution is strongly acidic. How could you make a strongly acidic solution neutral? Similarly, Mg^{2+} will precipitate with OH^- only if the solution is basic. How could you make sure a solution is basic? If the scheme you developed in Procedure 4 did not work, modify it along the lines suggested here and try it again.

QUALITATIVE ANALYSIS OF THE GROUP I IONS: Ag^+, Pb^{2+}, Hg_2^{2+}

OBJECTIVES

1. To become familiar with a standard qualitative analysis scheme.
2. To identify the ions present in a Group I unknown.

DISCUSSION

The general scheme of qualitative analysis will separate and identify a large number of positive ions. The standard scheme discussed in your text will handle 25 ions by separating them into four groups, I–IV. (These group designations are not the same as those of the periodic table.) Each group is separated in turn from a single sample by a group precipitating agent. The individual ions of the group are then further separated and identified.

The Group I ions, Ag^+, Pb^{2+}, and Hg_2^{2+} are separated from a general unknown by adding a hydrochloric acid solution. The precipitation reactions for the insoluble chlorides formed are:

$$Ag^+(aq) + Cl^-(aq) \rightarrow AgCl(s)$$

$$Pb^{2+}(aq) + 2Cl^-(aq) \rightarrow PbCl_2(s)$$

$$Hg_2^{2+}(aq) + 2Cl^-(aq) \rightarrow Hg_2Cl_2(s)$$

The flow chart below shows how the chloride precipitate is treated in five steps so as to separate and identify the individual ions.

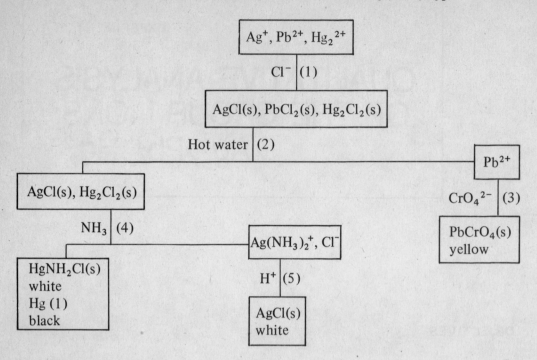

In this experiment we will first carry out the reactions that are used in the Group I qualitative analysis scheme. We will then use the scheme to analyze a Group I unknown.

PRELIMINARY STUDY

1. Review Sections 22.3 and 22.4 of the text.

PROCEDURE

Wear your safety goggles, apron, and gloves while performing this experiment.

PART I: Preliminary Tests

1. (a) Add 10 drops of 0.1 M $AgNO_3$ to a small test tube. Add 3 drops of 6 M HCl and note the character of the AgCl precipitate which is formed.
    (b) Centrifuge the AgCl mixture and pour off the liquid. Add 1 ml of 6 M NH_3 to the precipitate and stir well with a stirring rod. One product is $Ag(NH_3)_2^+$ ion.
    (c) Add 6 M HNO_3 dropwise to the above solution until it is acidic and note the results. The solution should be tested with litmus paper to make certain that it is acidic. Remember that acids will react with NH_3, even in a complex ion.
2. (a) Add 10 drops of 0.2 M $Pb(NO_3)_2$ to a small test tube. Add 3 drops of 6 M HCl and note the character of the $PbCl_2$ precipitate which is formed.
    (b) Centrifuge the $PbCl_2$ mixture and pour off the liquid. Construct a boiling

water bath in a 250 ml beaker. Add 1 ml of water to the precipitate and heat it in the boiling water bath for at least three minutes. Stir the mixture occasionally with a stirring rod.

(c) Add 2 drops of 1 M K_2CrO_4 to the warm solution from above. Note the character of the $PbCrO_4$ precipitate which is formed.

3. (a) Add 10 drops of 0.1 M $Hg_2(NO_3)_2$ to a small test tube. Add 3 drops of 6 M HCl and note the character of the Hg_2Cl_2 precipitate which is formed.

(b) Centrifuge the Hg_2Cl_2 mixture and pour off the liquid. Add 2 drops of 6 M NH_3 to the precipitate and note the results. The product is a precipitate of $HgNH_2Cl$ mixed with liquid mercury.

PART II: Analysis of a Group I Unknown

The unknown may contain 1, 2, or 3 of the Group I ions.

1. To 5 ml of your sample in a small test tube, add 0.5 ml 6 M HCl. Stir well and centrifuge. Pour off the liquid and discard it. Wash the precipitate with 4 ml water and 3 drops 6 M HCl. Stir well. Centrifuge and discard the wash liquid.

2. To the precipitate from Procedure 1, which contains the chlorides of the Group I ions, add about 4 ml water. Heat the test tube by putting it into the boiling water bath you used in PART I, Procedure 2(b). Keep it there for at least three minutes, stirring occasionally. Centrifuge quickly and decant the hot liquid, which may contain Pb^{2+}, into a test tube. Save any solid residue for Procedure 4.

3. Confirmation of the presence of lead.
    To the liquid from Procedure 2 add 3 or 4 drops of 1 M K_2CrO_4. The formation of a yellow precipitate of $PbCrO_4$ confirms the presence of lead. Centrifuge out the solid, which should be bright yellow.

4. Confirmation of the presence of mercury.
    If lead is present, wash the precipitate from Procedure 2 with 6 ml water in the boiling water bath. Centrifuge and test the liquid for Pb^{2+}. Continue washings until no positive reaction to the lead test is obtained. To the washed precipitate add 2 ml 6 M NH_3 and stir well. A black or dark grey precipitate establishes the presence of Hg_2^{2+}. Centrifuge and pour off the liquid, which may contain $Ag(NH_3)_2^+$ ion, into a test tube.

5. Confirmation of the presence of silver.
    To the liquid from Procedure 4 add 3 ml 6 M HNO_3. Check with litmus to see that the solution is acidic. A white precipitate of AgCl confirms the presence of silver.

SAMPLE DATA TABLE   **EXPERIMENT 35**

PART I: Preliminary Tests

Procedure   *Observations*

PART II: Analysis of a Group I Unknown

Procedure   *Observations*

Group I Unknown No. _____   Ions Present: _____

QUESTIONS

1. Write balanced chemical equations for each reaction which took place in the preliminary tests of PART I.
2. Describe each equation in Question 1, as best you can, as being associated with one or more of the following kinds of reactions: precipitation, solution, complex ion formation, complex ion decomposition, acid-base, oxidation-reduction (electron transfer).
3. Explain what could happen in the Group I analysis if the following mistakes were made:
    (a) hot water was used to wash the chloride precipitate in Procedure 1.
    (b) cold water was used in Procedure 2.
    (c) not enough HNO_3 was added in Procedure 5 to make the solution acidic.
4. Name a single reagent which will separate the following pairs of compounds:
    (a) $AgCl$, Hg_2Cl_2
    (b) $AgCl$, $PbCl_2$
    (c) $PbCl_2$, Hg_2Cl_2

EXTENSIONS

1. The qualitative analysis of Group I can be accomplished by many procedures. The one we use is best if other ions are also possibly present and need to be determined. Let us assume that we have a solution containing only one of the Group I ions. We wish to identify the ion by adding just one reagent to the solution. Find a reagent that will do the job. (There are actually several that might work, and the following list includes some of them: 6 M HCl, 3 M H_2SO_4, 6 M NH_3, 1 M K_2CrO_4, 6 M HNO_3, 6 M NaOH.)

ANALYSIS BY AN OXIDATION-REDUCTION TITRATION

OBJECTIVES

1. To carry out an oxidation-reduction reaction in solution.
2. To determine the strength of a household bleaching solution by means of an oxidation-reduction titration.
3. To gain experience with analytical procedures.

DISCUSSION

One of the important classes of chemical reactions is that of oxidation-reduction. In oxidation-reduction reactions there is always one species which is oxidized (whose oxidation number increases), and another which is reduced (whose oxidation number decreases). An oxidation-reduction reaction that is important in qualitative analysis is the following:

$$Sn^{2+}(aq) + 2Hg^{2+}(aq) + 2Cl^-(aq) \rightarrow Hg_2Cl_2(s) + Sn^{4+}(aq)$$

In this reaction the tin ion is oxidized, since its charge goes from +2 to +4. The mercury ion is reduced, since its oxidation number goes from +2 in Hg^{2+} to +1 in Hg_2Cl_2. Since Hg^{2+} ion oxidizes Sn^{2+} ion, the Hg^{2+} ion is the oxidizing agent in the reaction.

Oxidation-reduction reactions tend to go essentially to completion, and so are often used in analytical procedures, particularly titrations. In this experiment we are going to use an oxidation-reduction titration to analyze a household bleaching solution. Bleaching solutions (Chlorox, Purex, and so on) are used in laundering clothes. Similar chemicals are used to disinfect the water in swimming pools and water supplies. All of these reagents contain an effective oxidizing agent, a species capable of oxidizing another species. Chlorine, Cl_2, is the basis of most bleaching solutions and is used directly in the water purification systems in many cities. Bleaching solutions are usually made by dissolving Cl_2 in sodium hydroxide solution:

$$Cl_2(s) + 2OH^-(aq) \rightarrow OCl^-(aq) + Cl^-(aq) + H_2O \qquad (36.1)$$

The hypochlorite ion, OCl^-, contains chlorine in a +1 oxidation state; since the Cl^- ion is much more stable (−1 oxidation state), the OCl^- tends to be reduced in reactions, and so serves to oxidize other species. The hypochlorite ion readily oxidizes iodide ion in solution:

$$OCl^-(aq) + 2I^-(aq) + 2H^+(aq) \rightarrow I_2(aq) + Cl^-(aq) + H_2O \qquad (36.2)$$

Addition of hypochlorite ion to an acidic iodide solution causes the solution to turn brown as I_2 is formed. The iodine produced in the reaction can then be reduced back to iodide by thiosulfate ion:

$$I_2(aq) + 2S_2O_3{}^{2-}(aq) \rightarrow 2I^-(aq) + S_4O_6{}^{2-}(aq) \qquad (36.3)$$

In this reaction the color of the I_2 disappears as the I^- ion is re-formed.

Since the color of I_2 is easily seen, and since the presence of even small amounts of I_2 can be detected in the presence of a starch indicator, Reactions 36.2 and 36.3 offer an easy and accurate method for determining the hypochlorite content of a bleaching solution. One need only add iodide ion to a measured amount of bleaching solution and then titrate with a standardized solution of thiosulfate ion until the color of the iodine disappears. At that point all of the OCl^- ion will have reacted with I^- ion, and the I_2 produced will have been reduced back to colorless iodide ion. The number of moles of OCl^- will equal the number of moles of I_2 produced. Since it takes two moles of $S_2O_3{}^{2-}$ to reduce each mole of I_2, the number of moles of OCl^- in the sample will equal one-half the number of moles of $S_2O_3{}^{2-}$ added during the titration:

$$\text{no. moles } OCl^- \text{ ion} = 1/2 \text{ no. moles } S_2O_3{}^{2-} \text{ ion} \qquad (36.4)$$

Since both the hypochlorite and the thiosulfate solutions were made from sodium salts, Equation 36.4, when written in terms of those salts takes the form:

$$\text{no. moles } NaOCl = 1/2 \text{ no. moles } Na_2S_2O_3 \qquad (36.5)$$

In the experiment we will measure the volume of a $Na_2S_2O_3$ solution of known molarity required to titrate a measured volume of $NaOCl$ solution. Using these data we can calculate the number of moles of $Na_2S_2O_3$ used in the titration:

$$\text{no. moles } Na_2S_2O_3 = \text{Volume}_{Na_2S_2O_3} \times M_{Na_2S_2O_3} \qquad (36.6)$$

By Equation 36.5 we then obtain the number of moles $NaOCl$ in the sample. Knowing the molar mass of $NaOCl$, we can now find the number of grams of

NaOCl present. The percent NaOCl by weight follows immediately if we note that the density of the solution is just about 1.00 g/ml.

PRELIMINARY STUDY

1. Review Section 23.3 in the text.
2. (a) A student uses 24.0 ml of a 0.550 M $Na_2S_2O_3$ solution to titrate a 10.0 ml sample of bleach. How many moles of $S_2O_3{}^{2-}$ ion were used? (Ans.: 1.32×10^{-2})
   (b) How many moles of OCl^- ion were there in the sample? (Ans.: 0.660×10^{-2})
   (c) How many grams of NaOCl were there in the sample? What is the mass percent NaOCl in the bleach? (Ans.: 0.491 g, 4.91%)

PROCEDURE

Wear your safety goggles, apron, and gloves while performing this experiment.

1. (a) Add exactly 10 ml of bleach solution to a 250 ml Erlenmeyer flask. The bleach solution should preferably be measured with a pipet. Your instructor will show you the proper procedure. The solution must be pipetted using a suction bulb to draw up the liquid into the pipet.
   (b) Using graduated cylinders to measure volumes, add 20 ml of 1 M KI and 10 ml of 6 M HCl to the bleach. Mix well by swirling and note the color of the iodine which is produced.
2. (a) Rinse a clean buret with a few ml of a standardized solution of sodium thiosulfate. Fill the buret and the tip with the solution. Record the molarity of the solution and its initial level in the buret. Add thiosulfate gradually to the solution in the Erlenmeyer while swirling the flask, until the dark color of the solution begins to lighten. When the color of the solution is a light amber, add 2–3 drops of starch indicator. The solution will become dark blue due to the presence of iodine. Continue the titration carefully, drop by drop, until the solution becomes colorless. Record the final level of the $Na_2S_2O_3$ solution in the buret.
   (b) Repeat the titration with a second 10 ml sample of bleach. The volumes of $Na_2S_2O_3$ solutions used in the two titrations should be equal to within 1–2%.

SAMPLE DATA TABLE   **EXPERIMENT 36**

Molarity of $Na_2S_2O_3$ _____ M   Vol. of bleach _____ ml

	Trial 1	Trial 2
Vol. of $Na_2S_2O_3$ (initial)	_____ ml	_____ ml
Vol. of $Na_2S_2O_3$ (final)	_____ ml	_____ ml
Vol. of $Na_2S_2O_3$ used	_____ ml	_____ ml

CALCULATIONS AND QUESTIONS

1. (a) Calculate the number of moles of $Na_2S_2O_3$ used in the first titration (Eqn. 36.6).
   (b) Using Equation 36.5, calculate the number of moles of NaOCl in the bleach sample.
   (c) Given the molar mass of NaOCl, 74.5, now calculate the mass in grams of NaOCl in the sample.
   (d) Calculate the mass percent of NaOCl in the bleach sample. Assume that the density of the bleach solution is 1.00 g/cm^3.
   (e) Repeat the above calculations for the second titration. Average your final answers to determine the average mass percent of NaOCl.
2. Equations 36.2 and 36.3 are the overall equations for the oxidation-reduction reactions which take place in this experiment. Write the half-equations for each reaction. How many electrons are produced by each of the oxidation half-equations?

EXTENSIONS

1. Many solid household cleansers (Comet, Bab-O, Ajax) contain NaOCl as a bleach in addition to an abrasive and a detergent. Weigh out a 10 g sample of a cleanser and dissolve it in 50 ml of distilled water. (The abrasive will not dissolve.) Add only half as much of the KI and HCl solutions as used for the liquid bleach. Dilute your standardized $Na_2S_2O_3$ solution by a factor of ten to obtain the solution for this titration (10 ml $Na_2S_2O_3$ solution and 90 ml water, mix well). Titrate the sample using the same procedure as in the experiment. Calculate the mass percent of NaOCl in the cleanser.

OBJECTIVES

1. To learn the principles of voltaic cell design.
2. To construct some voltaic cells and measure their voltages.
3. To study the effect of concentration on cell voltage.

DISCUSSION

Oxidation-reduction reactions differ from other kinds of chemical reactions in that they are readily expressed as the sum of two half reactions, one involving an oxidation and one a reduction. If, for example, we put a piece of zinc metal into a solution of copper sulfate, the following oxidation-reduction reaction occurs:

$$Zn(s) + Cu^{2+}(aq) \rightarrow Zn^{2+}(aq) + Cu(s) \qquad (37.1)$$

The overall reaction can be written as the sum of the following half reactions:

$$Zn(s) \rightarrow Zn^{2+}(aq) + 2e^- \qquad (37.1a)$$

$$2e^- + Cu^{2+}(aq) \rightarrow Cu(s) \qquad (37.1b)$$

Ordinarily we write equations for half reactions mainly to simplify balancing oxidation-reduction equations. In the reaction as it occurs, when zinc is added to copper sulfate, we don't see any direct electron effects. Electron exchange must occur at the zinc surface, but if our theory didn't tell us we would never know that happens.

As with any reaction that proceeds spontaneously, Reaction 37.1 involves a substantial energy change that drives the reaction to the right. Unlike most other reactions, this one has an energy effect that can be tapped and made to do useful work. We tap the energy with a device called a voltaic cell. In a voltaic cell the oxidation half reaction occurs in one place and the reduction half reaction in another. The electrons released by the oxidation reaction flow in a circuit external to the cell, where they can do electrical work of one kind or another, and return to the cell at the point where the reduction reaction is going on.

Although the idea behind a voltaic cell seems reasonable, it is not at once apparent how to go about making such a cell. In Figure 37.1 we have shown a voltaic cell that works on the principles we have described and in which Reaction 37.1 can be made to do useful electrical work. The cell has two compartments separated by the porous bottom of the crucible. In the beaker we have a solution

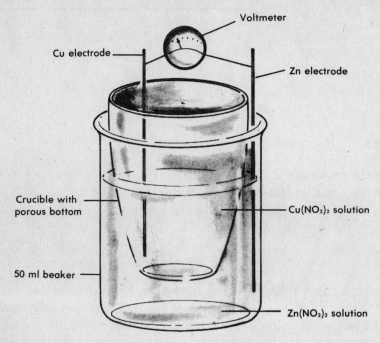

FIGURE 37-1   A voltaic cell.

of $Zn(NO_3)_2$ in which there is a zinc electrode, really just a piece of zinc metal. In the crucible we have a solution of $Cu(NO_3)_2$, in which there is a copper electrode. The half reactions 37.1a and 37.1b occur in the two compartments. Electrons from the zinc electrode, which is called the anode, pass through the external circuit, to the copper electrode, which is called the cathode. The electric circuit is completed, as it must be, by ions which carry charge through the porous bottom of the crucible. If we put a voltmeter across the two electrodes, we find the cell has a substantial voltage, about 1 volt.

The voltage of a cell like the one we have described can be expressed as the sum of a voltage associated with the electrode undergoing oxidation and a voltage at the electrode at which reduction is occurring. In the zinc-copper cell, the zinc electrode is oxidized and the copper electrode is the one at which reduction of Cu^{2+} occurs. For this cell:

$$E^0_{cell} = E^0_{oxidation} \; Zn \rightarrow Zn^{2+} + E^0_{reduction} \; Cu^{2+} \rightarrow Cu \qquad (37.2)$$

Given the voltage of the cell, and a value for E^0 at one of the electrodes, the value of E^0 at the other electrode can be calculated by Eqn. 37.2. Many electrode voltages have been determined this way, and they are listed in your text.

The voltage of a cell like that in Figure 37.1 varies with the concentrations of the cations in the two compartments. If you look at Reaction 37.1, you can see that, by Le Chatelier's principle, if we increase the Cu^{2+} ion concentration, we will tend to drive the reaction to the right. On that basis, we might predict that the voltage of the cell would increase if Cu^{2+} concentration went up, and indeed it does. If we decrease the concentration of Zn^{2+} ion we drive the reaction to the right and increase the cell voltage. Experimentally, it is easiest to decrease the concentrations of the two cations in this cell. To the zinc nitrate solution one might add 6 M NaOH, which will convert nearly all of the zinc to $Zn(OH)_4{}^{2-}$ complex ion, leaving almost no free Zn^{2+} in solution. If we add 6 M NH_3 to the copper nitrate solution, we will convert nearly all of the Cu^{2+} ion to $Cu(NH_3)_4{}^{2+}$ complex ion. Such changes have a large effect on the voltage of a zinc-copper cell.

A real disadvantage of the cell we have discussed so far is the fact that it needs two solutions, separated by a porous barrier. Such a barrier creates a large resistance in the electrical circuit, so that if we try to draw a substantial current in the cell, the cell voltage will fall rapidly, since so much work has to be done to get the electric charge through the solution. Any practical cell, one that will operate a flashlight or start a car engine, must have a low cell resistance. This means it must operate with only one solution.

It turns out that this condition, one solution only, is so restricting that there are really only a few practical voltaic cells. The main ones are the dry cell and the lead storage battery. A dry cell contains a zinc electrode and a graphite(carbon) electrode, both in contact with a moist paste consisting of a mixture of ammonium chloride and manganese dioxide. The cell has a voltage of about 1.5 volts. The electrode reactions appear to be:

$$Zn(s) \rightarrow Zn^{2+}(aq) + 2e^-$$

$$2e^- + 2\,MnO_2(s) + 8\,NH_4{}^+(aq) \rightarrow 2\,Mn^{3+}(aq) + 8\,NH_3(aq) + 4\,H_2O \quad (37.3)$$

The dry cell is useful for operating small radios and flashlights. It is not rechargeable and has relatively low capacity to deliver electrical energy.

At this point in time, the technological device that would be most useful to our society would be a compact, lightweight, high capacity, inexpensive, rechargeable voltaic cell. Such a cell could be used in electric automobiles. The availability of such vehicles would greatly decrease our need for gasoline. Scientists are working on the development of such a cell, but as yet we do not have one.

In this experiment we will assemble several voltaic cells. We will measure their voltages, determine the cell reactions, and observe the effect of concentration on cell voltage.

PRELIMINARY STUDY

1. Review Section 24.1 in the text.
2. Using electrode voltages in your text, find the standard voltage, E^0 cell, for the voltaic cell in Figure 37–1. (Ans. 1.10V)

PROCEDURE

PART I: Construction of a dry cell

Put about 5 g NH_4Cl and 5 g MnO_2 in a 50 ml beaker. Add a little water, enough to make a soft paste dispersed in a little liquid, and mix the two solids thoroughly with a stirring rod. Insert a zinc electrode and a graphite electrode into the paste. Measure the voltage of the cell by attaching the leads from a voltmeter to the two electrodes. If the voltage appears to be negative, exchange the leads on the electrodes.

PART II: Construction of voltaic cells

1. (a) Obtain a 50 ml beaker and a crucible with a porous bottom. Pour about 10 ml 0.1 M $Zn(NO_3)_2$ into the beaker. Put a clean zinc electrode into the solution. Pour about 10 ml 0.1 M $Cu(NO_3)_2$ into the crucible, and then put the crucible into the beaker. Put a clean copper electrode into the solution. Attach a voltmeter to the electrodes and measure the voltage of the cell. Which electrode is attached to the negative (−) terminal of the voltmeter? It is at the negative electrode that oxidation is occurring. The overall reaction that occurs is given by Equation 37.1.
    (b) Take the zinc electrode out of the zinc nitrate solution, rinse it with water and put it into the copper nitrate solution. Now measure the voltage between the copper and zinc electrodes. Examine the surface of the zinc electrode after it has been in the copper nitrate for a few minutes.
2. Make another cell like the one you assembled in Procedure 1(a), except this time use 0.1 M $Pb(NO_3)_2$ and a lead electrode in the crucible. Measure the voltage of the cell and note which electrode is negative. Make a third cell, this time putting 0.1 M $Cu(NO_3)_2$ and a copper electrode in the beaker, and 0.1 M $Pb(NO_3)_2$ and a lead electrode in the crucible. Measure the voltage of the cell and determine which electrode is negative.

PART III: The effect of concentration on cell voltage

1. Reassemble a zinc-copper cell like that used in Procedure 1(a) of PART II. Connect the voltmeter. To the copper nitrate solution add 6 M NH_3 until the solution in the crucible is dark blue and there is no precipitate. Record the voltage of the cell and the electrode that is negative.
2. Make a new zinc-copper cell, this time with the zinc nitrate solution in the crucible. With the voltmeter connected, add 6 M NaOH to the zinc nitrate solution until the voltage becomes steady. Record the voltage of the cell and the electrode that is negative.

```
┌─────────────────────────────────────────────────────────────────┐
│  SAMPLE DATA TABLE                            EXPERIMENT 37       │
│                                                                   │
│  PART I    Voltage of dry cell _____ volts                    │
│                                                                   │
│  PART II   Voltaic cell components   Voltage of cell   Negative electrode │
│                                                                   │
│            _____       _____ volts   _____   │
│                                                                   │
│  PART III  6 M NH₃ added to Cu²⁺     _____ volts    _____   │
│                                                                   │
│            6 M NaOH added to Zn²⁺    _____ volts    _____   │
└─────────────────────────────────────────────────────────────────┘
```

CALCULATIONS AND QUESTIONS

1. Explain why the dry cell constructed in PART I is useful as compared to the cells constructed in PART II.
2. (a) What are the cell reactions for the three cells you made in PART II? (Oxidation occurs at the negative electrode.)
    (b) Write the half reactions for each of the three cells.
3. (a) Express the voltage of the zinc-copper cell and the zinc-lead cell as the sum of an $E^0_{oxidation}$ and an $E^0_{reduction}$. Given that E^0_{ox} $Zn \rightarrow Zn^{2+}$ is 0.76 volts, find E^0_{red} for $Cu^{2+} \rightarrow Cu$ and for $Pb^{2+} \rightarrow Pb$.
    (b) Use the voltages you calculated in (a) to predict the voltage of the lead-copper cell you made in Part II. Does your prediction constitute a check on the theory? Explain.
4. Explain, as best you can, the change in voltage when the zinc electrode in the zinc-copper cell was put in the copper nitrate solution. What happened to the zinc electrode in the copper nitrate solution? What is the chemical reaction that occurs under those conditions?
5. Explain the effects on cell voltage in PART III when you added 6 M NH_3 and 6 M NaOH to the solutions in the zinc-copper cell.

EXTENSIONS

1. (a) In Procedure 1, PART III, the voltage changes by about 0.03 volts for each ten-fold decrease in the concentration of the Cu^{2+} ion. Using this information and the voltage data, estimate $[Cu^{2+}]$ in the solution after the NH_3 is added.
    (b) The dark blue solution contains Cu^{2+} ion, $Cu(NH_3)_4^{2+}$ ion, and NH_3 all in equilibrium. Assuming that the complex ion is 0.1 M and the NH_3 is 6 M, use the calculated value of $[Cu^{2+}]$ to find the equilibrium constant for the reaction: $Cu^{2+}(aq) + 4 NH_3(aq) \rightleftarrows Cu(NH_3)_4^{2+}(aq)$. It is from cells similar to this one that equilibrium constants for many reactions are determined.

WINNING A METAL FROM ITS ORE

OBJECTIVES

1. To simulate some processes that are used commercially to obtain a metal from its ore.
2. To determine the percent yield of lead from an ore.

DISCUSSION

One of the most important industrial processes is the extraction of a metal from its ore. In most cases the process begins with heterogeneous rock and ends with the recovery of a pure metal. Our industrial society consumes huge quantities of metal each year, nearly all of which are produced by chemical processes. In this experiment you will examine some of the processes used. The chemical and physical changes you will observe are the same as those used in the metallurgical industry.

The procedures used to win a metal from its ore are almost unique for each metal. Most processes, however, can be broken down into three steps: concentration, reduction, and refining. Metal ores are usually mixed with large amounts of rocky material from which they must first be separated. The "flotation" method of concentrating ore is based on the fact that oil floats on water and the fact that some minerals are more easily wet by oil than by water. The crushed mineral particles float to the surface with the oil when they are dispersed in an oil, water, and ore mixture. The bulk of the ore, mostly silica (SiO_2) and silicates, is wet by water and sinks to the bottom of the water layer.

Some ores can be reduced to the metal by "roasting", heating in air. The reduction of copper(I) sulfide at $1500°C$ is an example:

$$Cu_2S(s) + O_2(g) \rightarrow 2\ Cu(l) + SO_2(g) \tag{38.1}$$

Other sulfide ores, such as zinc sulfide, are converted to the oxide when roasted:

$$2\ ZnS(s) + 3\ O_2(g) \rightarrow 2\ ZnO(s) + 2\ SO_2(g) \tag{38.2}$$

Carbonate ores when roasted are decomposed to an oxide:

$$CuCO_3(s) \rightarrow CuO(s) + CO_2(g) \tag{38.3}$$

The oxides produced by roasting must then be reduced by a reducing agent. The most common reducing agent is carbon monoxide obtained from coke. The reaction of iron oxide with CO is typical:

$$Fe_2O_3(s) + 3\ CO(g) \rightarrow 2\ Fe(l) + 3\ CO_2(g) \tag{38.4}$$

The required purity of a metal varies with the metal and its intended use. The copper and aluminum used in electrical conductors must be at least 99.95% pure. Iron for cast iron is sufficiently pure after the reduction step in the blast furnace. There are several effective ways for refining metals; distillation, electrolysis, and special furnaces are a few of the ways used. In this experiment we will concentrate an ore by flotation and obtain a metal by reduction of an ore with charcoal.

PRELIMINARY STUDY

1. Review Chapter 25.
2. A bauxite ore contains 61% Al_2O_3. After concentration, reduction by electrolysis, and refining, 81% of the aluminum is recovered. Beginning with one kilogram of ore, how much aluminum metal would be recovered? (Ans: 0.26 kg)

PROCEDURE

Wear your safety goggles, apron, and gloves while performing this experiment.

The compounds in lead ores may occur as galena, PbS, white lead, $PbCO_3$, yellow lead, PbO, or red lead, Pb_3O_4. You will work with all of these compounds in this experiment.

Part I: Flotation of an ore.

1. Fill a large test tube half full with tap water. Add 2 grams of red lead ore (Pb_3O_4 and SiO_2) to the water. Stopper, shake thoroughly, and note the results. Now add mineral oil to the material in the tube so that its volume is about half that of the water. Shake thoroughly again and allow the test tube to stand. Describe what happened. Repeat the flotation procedure with a sample of galena ore (PbS and SiO_2).

Part II: Roasting and reduction of an ore.

1. Weigh out a 10 g sample of lead carbonate, $PbCO_3$, and put it in a crucible. (If the $PbCO_3$ is not finely powdered, grind it with a mortar and pestle.) Place the crucible on a clay triangle on an iron ring and heat with a bunsen flame. Continue heating until all of the material has changed color but avoid heating so intensely that the product is melted. Periodically remove the heat, and stir the contents of the crucible with a stirring rod to check on how the reaction is proceeding.

2. After the roasting reaction has been completed, empty the product into a mortar. Weigh out about 0.8 g of powdered charcoal and mix it thoroughly with the lead compound. Transfer the mixture back to the crucible and sprinkle a small amount of charcoal on the surface. Place the cover on the crucible and heat strongly. After about five minutes remove the cover and carefully stir the contents. Replace the cover and heat for another five minutes. Place the crucible and cover on a heat resistant pad.

3. Half-fill a 100 ml beaker with cold tap water. While the lead metal is still liquid, pick up the crucible with tongs and carefully pour its contents into the beaker. The lead will solidify as it is quenched by the water. If any lead remains behind, reheat the crucible and pour out any melt that forms. Decant the water from the beaker and wash the carbon off the lead with a stream of water from a wash bottle. Dry the lead on a piece of filter paper. Weigh the lead you obtained and record its mass.

SAMPLE DATA TABLE   **EXPERIMENT 38**

Process	*Observations*
Flotation	
Roasting	
Reduction	
mass of $PbCO_3$ _____ g	mass of Pb _____ g

CALCULATIONS AND QUESTIONS

1. (a) Calculate the theoretical amount of lead which could be recovered from the $PbCO_3$ sample used in the experiment.

   (b) Using the amount of lead actually recovered, calculate your percent yield.

2. Write the balanced chemical equations for the reactions that occur:

   (a) on roasting $PbCO_3$

   (b) on reduction of the product of (a) with carbon monoxide. Label the oxidizing agent and the reducing agent.

3. Lead can also be obtained from galena, PbS, by first roasting and then reduction with carbon monoxide. Write the balanced equations for these reactions. (See Eqns. 38.2 and 38.4).

EXTENSIONS

1. Beginning with copper carbonate, $CuCO_3$, extract copper metal using the procedures described in Part II. Write the balanced equations for the reactions and determine the percent yield of copper.
2. Metals are frequently recovered from solutions of their salts. Given a 0.1 M solution of copper sulfate, $CuSO_4$, devise a method by which the copper can be recovered. Use your text as a source of ideas.

EXPERIMENT 39

TRACE ANALYSIS
BY COLORIMETRY

OBJECTIVES

1. To develop a system for the determinations of small concentrations of ions.
2. To determine the limits of a method of trace analysis.

DISCUSSION

An important problem faced by chemists working on pollutants of the environment is that of analyzing for trace amounts of these materials. Concentrations of parts per million, or even parts per billion, may be sufficient to produce deleterious effects on plants and animals. There are many methods for trace analysis, and in this experiment we will examine one of the simpler ones.

One of the ways by which a species may be detected in solution is through its color. The color of a substance in solution varies in a predictable way with concentration. To detect small concentrations of any given species one usually treats the solution in which it is present with a reagent with which the species reacts to produce a highly colored product. One can measure the color by eye, by looking down a tube containing the colored product. This very simple method can be quite sensitive, since the eye is a remarkably effective light detector. By matching the color of an unknown with that of a standard solution one can determine the concentration of the species in the unknown.

In this experiment we will work with three ions: Fe^{3+}, Cu^{2+}, and Hg^{2+}. Each of these ions forms a highly colored substance when treated with a solution containing ferrocyanide ion, $Fe(CN)_6^{4-}$. Our purpose will be to determine the lowest concentration of each of these ions we can detect in a solution. Our method will be to make successive dilutions of the colored solution until we can no longer distinguish between the color of the solution and that of a sample of pure water.

PRELIMINARY STUDY

1. Review Section 21.3 in the text.

PROCEDURE

Students may work in pairs on this experiment. Wear your safety goggles, apron, and gloves while performing this experiment.

1. Pour 30 ml of a 1.10×10^{-3} M solution of $Fe(NO_3)_3$ into a clean, dry 100 ml beaker. To the solution add 3 ml 0.02 M $K_4Fe(CN)_6$. This will convert the Fe^{3+} to a blue-colored species; the dilution with the added reagent will bring the effective $[Fe^{3+}]$ in the solution to 1.00×10^{-3} M. This will be your stock solution for the analysis of Fe^{3+} ion. Pour the solution into a regular size test tube or flat bottom vial until it is 3/4 full. This container will serve as a viewing tube.

2. Measure out 10.0 ml of the colored stock solution with a 10 ml graduated cylinder and pour it into a 250 ml Erlenmeyer flask. Add 90 ml distilled water from a 100 ml graduated cylinder to the flask and swirl to thoroughly mix the reagents. Fill a viewing tube like that used in Procedure 1 3/4 full with this solution of 1.00×10^{-4} M Fe^{3+}. Look down the tube and check to see that the color is sufficient to detect.

3. Rinse out the 10 ml graduated cylinder with the diluted solution obtained in Procedure 2. Then use the cylinder to measure out 10.0 ml of that solution. Pour out the rest of the solution from the Erlenmeyer flask, and rinse the flask with water. Add the 10 ml of solution in the graduate to the flask. Then add 90 ml of water. Swirl to mix the reagents, and fill a third viewing tube 3/4 full of the resulting 1.00×10^{-5} M Fe^{3+} solution. Look down the tube to check that the color is still visible. If there is a doubt, fill two viewing tubes 3/4 full of distilled water and put them to the left and right of the sample. If you can distinguish the sample from the water, the color is visible.

4. Continue dilutions according to Procedure 3 until you can no longer see a difference between the diluted solution and water when both are at the same depth in the viewing tubes. The minimum concentration detectable lies between this solution and the last one in which the color was visible. Record the concentration of the last solution whose color you could detect.

5. Repeat Procedures 1 to 4, using a 1.10×10^{-3} M $Cu(NO_3)_2$ solution. Then repeat those Procedures with a 1.10×10^{-3} M $HgCl_2$ solution. With the $HgCl_2$, the color takes several minutes to develop.

SAMPLE DATA TABLE   **EXPERIMENT 39**

Cation tested	Solution color	Lowest concentration where color was detectable
Fe^{3+}	_____	_____ M
Cu^{2+}	_____	_____ M
Hg^{2+}	_____	_____ M

CALCULATIONS AND QUESTIONS

1. (a) What is the lowest molarity of Fe^{3+} that you were able to detect in this experiment?
   (b) How many grams of Fe^{3+} would there be in a liter of solution at that molarity? (The molar mass of Fe^{3+} ion is equal to the molar mass of Fe.)
   (c) What is the limit of detection of Fe^{3+} in grams per million grams (ppm) of solution by this method?
2. Repeat the calculations in Question 1 with Cu^{2+} and Hg^{2+}.
3. Suggest some ways that the general method used in this experiment might be made more sensitive.
4. The procedure used in this experiment has some drawbacks when applied to real samples. Suggest what some of the disadvantages might be.
5. Fish containing mercury at concentrations in excess of 0.5 parts per million are considered unsafe. Could the method used in this experiment detect as low a concentration as 0.1 parts per million of mercury in fish? Explain.

EXTENSIONS

1. Your instructor will furnish you a solution containing an unknown concentration of Fe^{3+}, Cu^{2+}, or Hg^{2+}. Given the identity of the ion, find its concentration in the solution by comparison with the standards you prepared. You will be able to improve the precision of your result by matching the color of your unknown with that of a particular dilution of the standard whose concentration is nearest to and greater than that of the unknown. See Experiment 21 for the procedure to use.
2. Absorption spectrophotometers can be used to measure the amount of light absorbed by a solution at a given wavelength. The maximum absorption of solutions prepared in this experiment occurs at about 650 nm for Fe^{3+}, 450 nm for Cu^{2+}, and 700 nm for Hg^{2+}. Use the spectrophotometer in your laboratory to determine the minimum detectable concentration of each of the cations studied. How do these concentrations compare to those obtained visually?

1	2

1 H 1.0079

3 Li 6.941 | **4 Be** 9.01218

11 Na 22.98977 | **12 Mg** 24.305

19 K 39.098	20 Ca 40.08	21 Sc 44.9559	22 Ti 47.90	23 V 50.9414	24 Cr 51.996	25 Mn 54.9380	26 Fe 55.847	27 Co 58.9332
37 Rb 85.4678	38 Sr 87.62	39 Y 88.9059	40 Zr 91.22	41 Nb 92.9064	42 Mo 95.94	43 Tc 98.9062	44 Ru 101.07	45 Rh 102.9055
55 Cs 132.9054	56 Ba 137.34	57 *La 138.9055	72 Hf 178.49	73 Ta 180.9479	74 W 183.85	75 Re 186.207	76 Os 190.2	77 Ir 192.22
87 Fr (223)	88 Ra 226.0254	89 'Ac (227)	104 § (260)	105 § (260)				

§The International Union for Pure and Applied Chemistry has not adopted official names or symbols for these elements.

★ Lathanoid Series

58 Ce 140.12	59 Pr 140.9077	60 Nd 144.24	61 Pm (147)	62 Sm 150.4

▼ Actinoid Series

90 Th 232.0381	91 Pa 231.0359	92 U 238.029	93 Np 237.0482	94 Pu (244)